VOLUME 5 ISSUE

Architectural Engineering and Design Management

Architectural Engineering and Design Management 5(1/2), April 2009

Published by Earthscan
2 Park Square, Milton Park, Abingdon, Oxon OX14 4R
711 Third Avenue, New York, NY 10017

Earthscan is an imprint of the Taylor & Francis Group, an informa business

ISSN 1745-2007 (print), 1752-7589 (online)
ISBN 13: 978-1-84407-895- 0 (pbk)

Abstracting services which cover this title:
Elsevier Scopus

Architectural Engineering and Design Management is published quarterly. Periodicals Postage Paid at Rahway, NJ. US agent: Mercury International, 365 Blair Road, Avenel, NJ 07001. POSTMASTER: Address changes to ARCHITECTURAL ENGINEERING AND DESIGN MANAGEMENT, 365 Blair Road, Avenel, NJ 07001.

Special Issue: Design Management for Sustainability

from Routledge

Guest Editorial: Design Management for Sustainability

Professor Stephen Emmitt

Building on the special edition *Aspects of Design Management* (Volume 3(1), 2007), this edition brings together the rapidly developing field of design management with that of environmental sustainability. Design management in the architectural, engineering and construction (AEC) sector should be an obvious vehicle for helping to bring about the implementation of environmentally sustainable ideals and practices. However, as reflected in the articles contained in this special edition, ranging from early policy decisions right through to facilities management, the reality is that there is work to do. This is partly to do with the emergent nature of design management within AEC, and partly to do with uncertainty and inconsistency in the application of environmentally sustainable regulations and tools. This appears to be an international phenomenon, as illustrated in the eight contributions from Australia, Denmark, South Africa, Sweden, the UK and the US.

In the first article, London and Cadman take a critical look at the fragmented regulatory environment in Australia and the influence it has on sustainable urban design management. Regulations, especially their interpretation and consistent application, are a fundamental issue for all development projects and the ability to manage design through the early development approval phases is an important task for design managers. In addition to providing useful insights into the Australian approval process, the authors also raise a number of generic issues concerning the role of the design manager in relation to sustainable urban development.

Continuing with the theme of early project stages, the public procurement of architectural services in Sweden is addressed in the second article. Sporrong and Bröchner review local procurement policies of 93 Swedish municipalities, revealing that very few have sustainability criteria in their tender assessments. Even where more general policies exist, they were reportedly not always followed. This article helps to highlight the need for better selection methods among public clients and tends to imply the need for design managers to persuade their clients to better reward sustainable design practices.

Corporate social responsibility is starting to attract the attention of researchers in the design management field and is a fundamental component of sustainable businesses. Othman provides an early insight into the perceptions of architects working in South Africa. Based on a series of interviews, the author provides some tentative suggestions for raising awareness and supporting corporate social responsibility. Hopefully this article will form the catalyst to more research in this under-researched area of business.

In the fourth article, Peat investigates how materials and products with sustainable credentials are marketed to specifiers. Scrutinizing advertisements in a select number of publications helped to reveal a number of shortcomings with how green products were promoted, which could possibly hinder their uptake. These findings have parallels with some of the early research into the specification process which also called for better-quality information from manufacturers. It is to be hoped that marketing departments will take heed of these findings and start to provide specifiers and design managers with the information required to help advance an environmentally sustainable agenda.

Returning to Scandinavia for the fifth article, Nielsen *et al* discuss the courage and competences required to implement sustainable housing in

ARCHITECTURAL ENGINEERING AND DESIGN MANAGEMENT ■ 2009 ■ VOLUME 5 ■ PAGES 3–4
doi:10.3763/aedm.2009.0901 © 2009 Earthscan   ISSN: 1745-2007 (print), 1752-7589 (online)

Denmark, a country long regarded as a frontrunner in sustainable housing. The authors' research reveals a widespread resistance to implementing sustainable measures, but encouragingly, also highlights the importance of networks and innovators in helping to overcome the barriers. Inspired by network theory, the research findings help to shed some light on the complex interaction of project participants and the role of authorities. Their findings should be useful in helping design managers to plan projects strategically to maximize opportunities for sustainable building.

As intimated in most of the articles included in this special edition, the design process for sustainable buildings appears to be vaguely defined. In the sixth contribution, Magent *et al* report on the development of a technique to model and evaluate the design process. This is an area in which many new methods, techniques and models are starting to emerge and be promoted, often without due regard for the practicalities of practice. The authors of this article have set out to provide practical guidance to designers grounded in research into sustainable buildings in the US. Their research raises a question about whether processes for sustainable buildings should be different from those employed for less sustainable buildings, and if so, to what extent?

In the seventh article, the attention shifts to the role of the construction design manager. In the UK, contractors are playing an increasingly dominant role in design decisions through design and build procurement routes, with the major contractors starting to employ a significant number of construction design managers. Thus a natural line of inquiry is to look at the potential contribution of the construction design manger to the delivery of sustainable buildings. Mills and Glass found a lack of clarity in the understanding and application of design management, consistent with earlier work, together with a series of barriers to sustainable development. By focusing on skills acquisition and skills improvement a number of recommendations are made.

Facilities management forms an integral aspect of design management in AEC. In the final contribution the authors provide a number of insights into the developing role of the facilities manager and their attitudes towards environmental sustainability. Elmualim *et al* report on a questionnaire survey into facility managers' perceptions of, and commitment to, sustainable issues, concluding that there is a significant knowledge gap. Given the facility manager's responsibility for running the buildings and their potential contribution to the client briefing process, these results give rise for concern. It would appear, as also evidenced in some of the other papers, that the knowledge of sustainable issues varies considerably between participants in a project; something that design managers need to be aware of.

There can be little doubt that the processes by which people make, use and remake buildings will have an impact on our planet. Design management, regardless of how it is perceived and applied, can and must make a positive impact on our activities. Reflecting on the contributions to this special edition it is evident that there are many positives to be drawn from the papers, but it is also evident that there are many difficulties to overcome and much work to do. As a community we must strive to bring about effective and environmentally aware design management practices throughout the entire life cycle of projects. How we do this in an effective, efficient and ethical manner, as witnessed in the articles, is still very much open to question.

On a personal note, I would like to dedicate this special edition to the memory of Dr David P. Wyatt, who passed away in 2008. David will be known to many of our readers both for his work on the ISO series 15686 *Buildings and Constructed Assets – Service Life Planning* and for raising the awareness of many (myself included) in the AEC sector to environmental issues, especially within the CIB's W096 Architectural Management community. Although some of the terminology and thinking advocated by David has started to enter the mainstream, our collective challenge continues to be to change our habits and become responsible custodians of our natural and built environments. Hopefully the contents of this special edition will go some way towards stimulating that objective within the design management discipline.

ARTICLE

Impact of a Fragmented Regulatory Environment on Sustainable Urban Development Design Management

Kerry Anne London and Katie Cadman

Abstract

The building project development approval process is increasingly complex and fraught with conflict due to the rise of the sustainable urban development movement and inclusive decision making. Coupled with this, government decision-making decentralization has resulted in a fragmented and over-regulated compliance system. Problems arising from the process include wasted resources, excessive time delays, increased holding and litigation costs, inadequate planning coordination, high levels of advocacy costs and a divisive politicized approval process. In Australia, despite attempts by government and industry associations, numerous problems are still unresolved. Design managers increasingly assume a liaison role during the approval phase. There is a long tradition of planning theory literature which provides context for understanding the knowledge–power–participation relationship for this paper. This study investigated the policy, process and practice conflicts during the approval stage in achieving sustainable urban developments. Three regional local government areas within one state jurisdiction and observations from detailed structured focus group interviews involving 23 stakeholders, proposers and assessors were analysed to explore this conflictual environment. As a result of regulatory fragmentation and excessive consultation, various persuasion tactics have been developed by all stakeholders of which 'reciprocity' and 'authority' were identified as the most common. Two challenges for design managers were thus identified: first, the emergence of the role of a *by default* central informal *arbitrator* across conflicting planning instruments; and, second, as a *navigator* through a set of persuasion tactics. An inclusive knowledge-based design management framework for sustainable urban development is proposed considering Habermas' communicative planning theory, Foucaltian governance and discursive powers thesis and Cialdini's persuasion theory, as well as being grounded in the key empirical results from this study, using various types and sources of knowledge as an authoritative persuasion tactic.

■ *Keywords* – Building design management; development approval; sustainable urban development; authoritative persuasion; inclusivity

INTRODUCTION

For sustainable urban development, a complex and fragmented regulatory planning system has become an intrinsic feature of the development application process which can produce a negative conflictual environment if not managed well (PCA, 2006, 2007).

For the successful implementation of sustainable urban development policies, effective coordination of policy action at all levels of government, including regional, state and federal, is required (Innes, 1993). Errors in planning instruments are often identified and never seem to be resolved (McKenzie, 1997). Numerous conflicts appear to occur because of inconsistencies that have arisen within jurisdictions and at times across jurisdictions in relation to the various planning instruments and the various

ARCHITECTURAL ENGINEERING AND DESIGN MANAGEMENT ■ 2009 ■ VOLUME 5 ■ PAGES 5–23
doi:10.3763/aedm.2009.0902 © 2009 Earthscan ISSN: 1745-2007 (print), 1752-7589 (online)

from Routledge

interpretations related to sustainable urban development (Innes, 1993).

It is speculated that coordination mechanisms for policies between multiple layers of government are often weak or non-existent and consistency between state and local land development plans is typically lacking (Peters, 1998; Weitz, 1999). Some form of regional governance and coordination is needed to transcend local boundaries and serve as a bridge between local communities and state government (Bengston et al, 2003).

The impacts of the increasingly conflictual environment have been widely reported and include increased uncertainties, lengthy protracted community consultative planning processes (Mayer, 1997), political and financial risks, corruption, divided communities, excessive time delays and associated holding costs, increased regulation compliance costs, a litigious environment and increased cost of consultant reporting (PCA, 2006, 2007).

PLANNING REFORMS

In Australia, the conflict between local and state governments, and from one jurisdiction to another, is of significant concern for the development industry and presents challenges for design managers. For development compliance there is a complex and fragmented system of numerous planning instruments including state Acts and state and local planning controls; on a project there can be more than 20 different planning instruments that require detailed knowledge and attention. The variances alone between states and between local jurisdictions has resulted in an increasingly complex and disharmonized system which has a significant impact upon a market of suppliers who increasingly work across jurisdictions; without the added frustration of conflicts within the 'set' of instruments related directly to a project (Williams, 2007).

An 'industry story' of the attempts to solve problems associated with the Australian development approval system can be traced to the mid 1990s (DAF, 1996; Australian Government Productivity Commission, 2004; DAF, 2005, 2006; London and Chen, 2007). Despite a series of industry investigations, national studies, reports and planning reform initiatives, there has been a general

acceptance that the planning approval process in some places in Australia is still a system in crisis, lurching from one reform to another.

In 2003, a longitudinal study conducted by the Royal Australian Institute of Architects (RAIA, 2003) involved distributing surveys to a sample group of 2026 architects across Australia. The study was repeated in 2005 with the same sample group (RAIA, 2005a). Although the study did not indicate response rates, the RAIA membership is approximately 9000. While results indicated some minor improvements to the Development Approval (DA) process in some states, the results suggest that any improvement had been 'insufficient to offset rising time delays and compliance costs across the country' (RAIA, 2005b). Average times for approvals reportedly increased in almost every state between the two survey periods. In many cases, the increases were far from minor with approval times in some states doubling.

Various state governments have made serious attempts to solve the problem, for example in NSW in 2006 a state planning reform was introduced. The primary change was a shift from local government control of the approval process for major projects to state government control. In early 2008, major planning reforms in the state of New South Wales (NSW) were again proposed to address the consequences of the ill-fated planning reforms implemented in early 2006. The 2006 reforms simply did not solve the problems and may have in some cases exacerbated the situation. The underlying premise in the 2006 reforms was that state planning could process more effectively and efficiently, and there was no evidence to indicate that this occurred.

While acknowledging the recent reforms' attempt to solve procedural problems, there is still considerable widespread criticism regarding the latest round of proposed reforms, ranging from claims that the reforms are insufficient to concerns about the reduction of community consultation (UTA, 2008; EDO, 2008; LGSA, 2008). It has been a long-held belief that local governments have a significant role to play in achieving sustainable urban development and this extends back to the 1992 Earth Summit (Robinson and Edwards, 2009). Internationally there have been some measures of success with the high engagement of local governments with the Agenda 21 programmes (Tibaijuka, 2002), actively involving some 6000 local governments in 100 countries. Apparently in Australia there has been little consideration given to issues of sustainable urban design at the local government level (Gurran, 2002).

COLLABORATIVE PLANNING PROCESSES

The past two decades have seen the rise of a highly participatory planning process. Initially aimed at improving democratic and transparent planning processes, it appears that other problems have beset this environment. The real contribution of the participatory planning process has also been called into question as it is claimed that in many cases the participatory planning process does little more than grant legitimacy to 'selling' the proposal (Lyle, 1985; Mayer, 1997; Tewdwr-Jones and Thomas, 1998; Steiner, 1999). There is an ever-increasing desire to challenge the claims made by proponents of development proposals in relation to sustainability by various local community groups. Consultative processes have increased in the past decade and to achieve timely development approvals, many involved in the process are required to persuade community stakeholders and various planning authorities about the extent that the proposal achieves sustainability across economic, social and environmental objectives. It is suspected that the increased demand to solve and mediate disparate views on sustainability has emerged as one of the most significant issues for design managers today. The skills required to meet social, economic and environmental objectives of the various planning instruments and to present professional and scientific arguments to support design solutions is perhaps one of the most challenging aspects of design management for the 21st century.

Various collaborative decision-making models have been considered and developed to attempt to integrate stakeholders' and disciplinary knowledge and/or diverse objectives, develop a shared understanding of the diverse interests and create solutions (Mayer and Seijdel, 2005; Golobic and Marusic, 2007) that are typically aimed at conflict resolution. The desire to produce normative, theoretical-based models of integration and collaborative planning practice tools and exhort the success of such tools does little to question the underlying practical difficulties that beset this phase of project development. This has been brought into question by Watson (2003) as he states '... the reality of fundamentally different world views and different value systems is still often treated as superficial in planning theory...'. He also calls for '... a return to the concrete, to the empirical and to case research ... as a way of gaining better understanding of the nature of difference, and generating ideas and propositions which can more adequately inform practice'.

There are limited examples of these types of empirical studies that examine the fundamental characteristics of the conflictual environment during the development approval phase, the underlying causes and then the negotiation practices to reduce conflict in an increasingly complex and fragmented planning system. In addition, there are studies that assist planners; however, there are few that assist in understanding the role of the design manager in the development approval process. The paper explores interpretations of sustainable urban development from community stakeholders', project proposers' and project assessors' perspectives, the key sources of conflict and an evaluation of persuasion tactics undertaken to influence others with respect to design management.

DESIGN MANAGEMENT: CONFLICT RESOLUTION

Multi-stakeholder projects have ensured that the design manager is crucial to the acceptance, progress and completion of the project by the community, client and regulatory authorities (Anderson et al, 2005). The role of design manager in the construction industry continues to broaden and diversify. The role of design management has been constantly evolving since the 1960s as design managers '... have sought to better understand the design process and the interdependency of organisations and individuals contributing to construction projects' (Emmitt, 2007). There are various project team members who have a role in managing an aspect of design (Gray and Hughes, 2001) and thus various definitions of who is the design manager. However, for the purposes of this project, we considered the design manager as the person responsible for coordinating the design of the project in the early initiation stages and who takes responsibility for the development approval. There are many challenges for the design manager in the early phases including; integration of design consultants, brief development, client and user liaison, community consultation and liaison with various authorities. Projects are nearly always political in nature and to varying degrees members of the community can play a role in shaping the direction of projects.

There is an increasing need to gain consensus with the various disparate external groups outside the core design team in relation to the sustainability qualities of a project. Interestingly, consideration of these 'others' is often in a grouped manner as they are considered a homogenous group; with the implication that they act similarly. The collaborative movement has focused on giving a voice to disenfranchised and/or powerless groups such as 'users' and the 'community'. However, we still tend to treat the users, community stakeholders, design consultants and regulating authorities as speaking with one voice (Figure 1) and yet one suspects the complexity of different types of groups has increased and that they are more heterogeneous than we have anticipated – something akin to Figure 2. It is speculated that the various government agencies have little cohesion at all. The failure of existing complex planning processes to address sustainability in a coordinated and holistic manner is part of the increasingly important role of design management in the construction industry (Anderson et al, 2005).

The need for a high level of integration of disciplinary insights and stakeholder perspectives in relation to sustainable urban development is not particularly new (Mayer and Seijdel, 2005). The collaborative and participatory process to sustainable urban development planning and decision making has focused on stakeholder management and the plan approval authorities' role (Mayer and Seijdel, 2005) whereas there is little theoretical and empirical work that provides a framework for design management. Sustainable urban development planning decisions are the responsibility of government and planning authorities; however, sustainable urban development design management for particular projects is the responsibility of the design manager.

The use of methods to influence the various interested actors to believe and/or act in certain ways regarding a project proposal can be varied. Drawing from social and organizational psychology literature it is apparent there are several influence methods that have been described (Cialdini, 1993; Yukl, 1998; Greene and Elfrers, 1999; Higgins et al, 2003).

Yukl (1998) describes influencing behaviours in relation to power and defines three sources of power – position, person and expert. First, 'position power' which is derived from legitimate power invested through statutory or organizational authority control over rewards/punishments/information; second, 'personal power' which is derived from human relationship influences or traits including friendship/loyalty and charisma; and third, 'expert power' which is a function of a leader's relatively greater knowledge about the tasks at hand when compared with subordinates who are dependent on that knowledge.

Greene and Elfrers (1999) outlined several forms of power including coercive, connection, reward, legitimate, referent, information and expert which seem to provide more detail to Yukl's descriptions. Coercive influencing tactics are based upon fear where the failure to comply results in punishment and relates to positional power. Connection refers to power resulting from connections to networks of people with influence and is both personal and political power. Reward is based upon the ability to provide rewards through incentives to comply and is typically related to position power. Legitimate is based on organizational or hierarchical position and is related to both position and political power. Referent is based on personality traits such as being likeable and admired and is personal power. Information is based on possession of or access to information perceived as valuable and can be a combination of position, personal and political power. Expert is based upon expertise, skill and knowledge and is related to an influencing power vested in a person and, if credible, the respect influences others and is considered to be personal power.

Cialdini's (1993) persuasion theory may also provide insights when considering tactics in relation to conflict resolution during the development approval process. Cialdini (1993) is the most cited

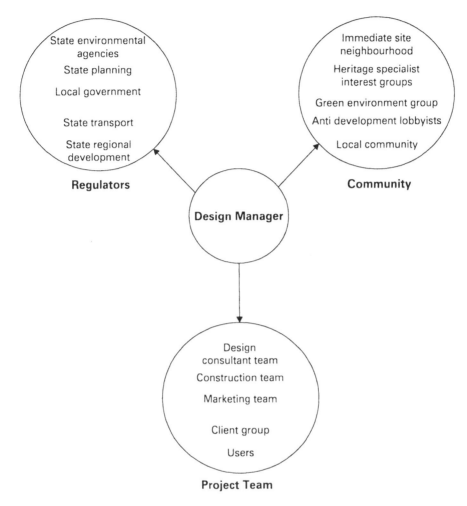

FIGURE 1 Tripartite model of influencing groups for sustainable urban development

social psychologist who has identified six key principles of persuasion through a study of the psychological principles that influence the tendency of professionals to comply with requests for change. The six key principles are:

- reciprocity
- social validation
- commitment and consistency
- friendship and liking
- scarcity
- authority.

'Reciprocity' is where one should be more willing to comply with a request from someone who has previously provided a favour and concession. 'Social validation' is where one should be more willing to comply with a request or behaviour if it is consistent with what similar others are thinking or doing, i.e. we may tend to echo the behaviours of others for various reasons. 'Commitment and consistency' is where after committing to a position, a person is more willing to comply with requests for behaviours that are consistent with that position. 'Friendship and liking' refers to the situation where one should be more willing to comply with the requests of friends or other liked individuals. 'Scarcity' refers to situations where one should try to secure those opportunities that are scarce or dwindling, i.e. if people perceive they are going to miss out on something they are

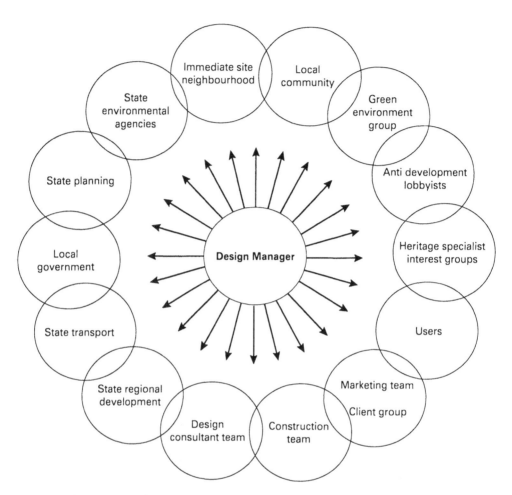

FIGURE 2 Specialization and fragmentation of influencing groups for sustainable urban development

more likely to be persuaded to change their behaviours. 'Authority' is where one should be more willing to follow the suggestions or instructions of someone who is a legitimate authority.

This general discussion on persuasion tactics is useful when considering the role of the design manager and in particular in the situations where a design manager is required to influence stakeholders, clients and/or regulating authorities. From a rationalist's perspective, clear prescriptive policies, planning instruments, guidelines, tools, qualitative information, scientific evidence, precedences and logical arguments would tend to appeal and persuade and one would assume this is the approach taken by most professionals. However, appealing to emotional responses and values and ideals is a valuable tactic along with encouraging commitment to projects through consultative, participative and collaborative approaches. The following section outlines the methodology undertaken to explore empirically the nature of conflict in relation to sustainable urban development and the approval system and the collective response from stakeholders, development proposers and development assessors to conflict experiences.

METHODOLOGY

The fieldwork was limited to NSW, the most densely developed state in Australia with the highest population. While the study is limited to a particular geographical region, design managers in other states of Australia, and indeed other countries, experience many of the problems noted. This paper reports part of a larger study whose research objectives were to identify the barriers and drivers for sustainable urban development decision making in regional Australia.

A case study methodology was deployed to investigate three local government areas consisting of two to four councils. Eleven semi-structured focus group interviews (FGIs) were conducted across the three case studies designed to canvass the interpretations and experiences of 23 participants (Table 1). FGIs duration ranged from 2–4 hours and involved asking questions in three main areas:

- interpretations of sustainable urban development
- experiences of conflicts in development processes
- ideas for solutions.

The interviews were divided into the following groupings – stakeholders in the community, proposers including architects/clients/developers and assessors including local government agencies.

There were three stages in the data collection and analysis:

- Stage 1 – Raw data collection: taping interviews and/or collecting documents, etc.
- Stage 2 – Data reduction: developing transcripts and coding documents/transcripts for common themes.

TABLE 1 Overview of participants interviewed for the research

CASE	ORGANIZATION/ GROUP	NO OF PARTICIPANTS
1 Sydney Surrounds	Stakeholder	3
	Proposer	1
	Assessor	1
2 Mid North Coast	Stakeholder	4
	Proposer	4
3 Lower Hunter	Stakeholder	3
	Proposer	5
	Assessor	2
Total		23

- Stage 3 – Data analysis and displays: interpreting coded data and relating it to the constructs/ themes and producing matrices that summarize the key concepts.

Stage 3 involved a two-part thematic analysis. The first step involved 'open coding' whereby the association of themes and concepts were revealed by the individual transcripts and eventually both a within-case and cross-case analysis was conducted and the comparative analysis presented in matrices. Step 2 involved 'axial coding' which is the rearrangement of data according to dominant themes that have emerged.

RESULTS AND DISCUSSION

Transcripts were coded and a two-stage thematic analysis was completed. First, an identification of the key themes was carried out in relation to:

- interpretations of sustainable urban development, and
- sources of conflict.

The second stage of thematic analysis involved:

- a discussion of key themes of impact of regulatory fragmentation and persuasion tactics, and
- an evaluation of persuasion tactics in relation to theory and role of design management .

INTERPRETATION OF SUSTAINABLE URBAN DEVELOPMENT

Participants from three groups from the three case study regional areas were asked to define and discuss their interpretations of sustainable urban development in focus group interviews. The results highlighted the diverse range of interpretations. It became apparent that this confusion was often ultimately at the 'heart' of many of the conflicts that were discussed during the interviews and that persuasion tactics can be traced to definitions and an individual's stance on the sustainable urban development concept.

There was a high degree of variation both within and across the cases regarding the participants' interpretation of sustainable urban development (Table 2). The themes arising commonly fall into four main areas:

TABLE 2 Participants' understanding of sustainable urban development

	CASE 1	GROUP	CASE 2	GROUP	CASE 3	GROUP	
Balance of various elements	Environment, social, infrastructure	S	Environment, economic – consistent	P A	Economic, social and environmental, social, natural, human and economic	P A	5
Focus/priority on one element	Environment, social	P A	Environment	S	Social	P	4
Maintenance of existing amenity	Minimal impact on environment	S	Social character; not over-developed	S P	Heritage, geography, ecology	S	4
Subjective		S P A	Manipulated by individual interests	S P A	Case-by-case basis, elusive/unclear	S P A	9

S – Stakeholder; P – Proposer; A – Assessor

- a balance of various key elements including economic, social, environmental, infrastructure
- a focus on one key element, for example, environment or social objectives
- the maintenance of existing features/ characteristics of an area such as its social character or environmental condition
- a subjective matter and thus largely difficult to define.

In all three case studies, the stakeholders described sustainable urban development as the capacity to maintain the existing desirable components and amenity with distinctive diverse subthemes such as sustaining the social character and protecting the heritage, geography and ecology. The first case study stakeholders were all quite adamant that sustainability was only about the environment and that other interest groups – developers, architects and authorities alike – had 'hijacked' the sustainability argument to include their own economic and/or social objectives. So there were clear diverse interests and complex arguments and undercurrents within the stakeholder group. This is, however, the only underlying broad theme that was consistent across the three regional case studies for a specific group.

Further to this, the stakeholder groups typically felt that proposers and assessors alike lacked local knowledge of the historical, cultural, social and geographical context of a specific region. The cultural practices and the cultural importance of the way of life in the region, as well as the maintenance of existing scale and character of localities, was seen to be eroded by what the stakeholders perceived to be unsustainable development in which local knowledge was ill-considered, unknown and/or undervalued. This has implications for persuasion tactics and moves beyond the need for community stakeholder groups to simply be included in consultation and refers to the legitimacy provided by authoritative persuasion using substantive knowledge with rational arguments based upon information, factual evidence and historical insights. It was typically claimed that the substantive knowledge was located in the collective memories of the community and so there is still merit in the principles underpinning power to influence people through human relationships, social connections and networks.

The knowledge of the various policies, planning instruments, Acts and processes towards achieving sustainability within selected parts of the stakeholder community groups was substantial. The knowledge networks between different regional stakeholder groups were also substantial and had become much more formalized in recent years. Transcripts of court cases were referred to and were subsequently checked by the researchers which provided evidence that cohesive and logical arguments in relation to design were engaged in between members of the judiciary, developers and architects. In some cases, the defence by professionals was somewhat lacking.

Underpinning the stakeholder group's view of sustainable urban development was a strong 'anti-development' approach to development. All stakeholder participants were able to offer previous accounts in which their objections had been central to the prevention of various proposed development projects. Strong mobilization by resident action or special interest groups has contributed to many conflicts within the development approval environment.

Apart from this one instance, there was no real consensus about the term 'sustainable urban development' between the participant groups. It is of concern that the key players who are involved with and contribute to the achievement of sustainable urban development do not have an agreed understanding of sustainable urban development. Perhaps more concerning is that the only time all participant groups were in agreement was when describing sustainable urban development as a subjective and a highly contentious matter. Case 2 and 3 participants viewed sustainable urban development as 'elusive', 'diluted' and 'bastardized to suit any political or personal ends depending on your agenda'. The consensus was that it was not typically clearly articulated in planning documents including state, regional and local environmental plans. To be more specific, there was diversity in scope, detail and approach; for example, in some planning instruments, sustainability objectives were quite prescriptive, whereas the same type of plan in a local government nearby would not be prescriptive and the documents and approaches taken were ever-changing and interpretative. In these localities, the method of assessment often relied upon expert reports to substantiate anticipated achievement of sustainable urban development. The ever-changing expectations referred to by architects and developers in these situations caused frustrations and it was considered by the proposers that assessors' understanding of sustainable urban development was not objective and measurable.

SOURCES OF CONFLICTS

The next stage of analysis was in response to the question of where the source of conflicts arose which led naturally on from the discussion about defining sustainability. This revealed an array of conflicts that have been categorized as policy, process or practice issues in the summary in Table 3.

There were two key sources of conflict. First, categorized as a policy issue, conflicts are a result of

the high level of inconsistency between regulations across the three scenarios; local–local, state–state and state–local. The regulatory environment has evolved as a mechanism to control and define sustainable urban development with the aim to reduce potential conflict; yet, the number and fragmentation of regulations has increased. This in turn has given rise to greater potential for conflict to arise between various policies, plans, codes and regulations. The increased knowledge of the stakeholder community of the detail within planning instruments allowed for a number of these participants to establish with authority in the interviews their views on the inconsistency that had arisen between regulations as they had experienced this through the judicial system. Another part of this study examined particular concrete examples of inconsistencies where different agencies responsible for different aspects of project approval have conflicting objectives, statements and conditions and, in each case, the design manager has had the task of negotiating between the various agencies to achieve some form of consensus.

The second main theme identified was that in practice, values, attitudes and personal relationships had become much more important in the development approval process and that now there was a high level of subjectivity. There has emerged an environment of persuasion where each participant is attempting to persuade other participants in the process to believe that the development is or is not achieving certain sustainability objectives. This excessive consultation between many parties appears to be a way forward to resolving conflicts and explaining the relative merits of the development proposed which cannot be so clearly articulated through objective criteria in any regulatory documentation (i.e. codes, plans, policies, etc.). This was intrinsically tied to policy, process and practice, namely that there was a high level of agreement between all groups that the subjective interpretation of planning instruments and interpretation of sustainability in relation to a development relied upon individual attitudes and personalities involved and this caused or reduced conflicts and disputes within the procedural matters in relation to approval. The manner in which projects are presented in the media, to

TABLE 3 Summary of key themes identified in relation to policy, process and practice conflicts

	CASE 1	GROUP	CASE 2	GROUP	CASE 3	GROUP	
Policy	Inconsistent regulations: local–local, state–state, state–local	S P A	Inconsistent regulations: local–local, state–state, state–local	S P A	Inconsistent regulations: local–local, state–state, state–local	S P A	9
	Increased volume and complexity	P	Increased volume and complexity	P A	Increased volume and complexity	P A	5
Process	Single point assessment issues	S	Single point assessment issues	P A	Single point assessment issues	S P A	6
	Local community appeal process issues	A	Local community appeal process issues	P	Local community appeal process issues	A	3
	Fragmented approval process (state–local, state–state, local–local)	P A	Fragmented approval process (state–local, state–state, local–local)	P	Fragmented approval process (state–state, local–local)	P	4
			Council officers lack empowerment	P	Council officers lack empowerment	P	2
Practice	High level of political support – relationships	S P	High level of political support – relationships	S P	High level of political support – relationships	S	5
	Subjectivity: attitudes, personalities	S P A	Subjectivity: attitudes, personalities	S P A	Subjectivity: attitudes, personalities	S P A	9
			Lack of skilled resources – state and local	S P	Lack of skilled resources	P A	4

S – Stakeholder; P – Proposer; A – Assessor

community forums, to assessment panels, was part of an emerging culture of negotiation and persuasion as to what constituted sustainable urban development.

The two key themes of the increasingly inconsistent, fragmented and complex regulatory environment experienced by participants and the emergence of an environment of persuasion are now discussed in more detail.

REGULATORY ENVIRONMENT

In NSW, planning and development is carried out under the Environmental Planning and Assessment Act 1979 and Environmental Planning and Assessment Regulation 2000. The Act and Regulation document the type of development that requires development approval and the processes for approval. Within the planning system there is a hierarchy of environmental planning instruments including state environment planning policies (SEPPs), regional environmental plans (REPs) and local environmental

plans (LEPs). State, regional and local plans indicate the level of assessment that is required and by whom: council, an accredited private professional or the Minister for Planning. Further to the LEPs there are also development control plans (DCPs), which are used to help achieve the objectives of the local plan by providing specific, comprehensive requirements for certain types of development or locations. Any particular site can have more than one DCP although there are steps towards rationalizing DCPs with the current round of proposed planning reforms. Prior to the 2006 reforms, the local government agency assessed all projects within its locale and then post-2006, the Minister for Planning through the state planning agency assessed and approved projects over A$50 million and/or projects of state significance.

There was a continuing discourse regarding the various planning instruments and the increasingly complex regulatory environment. Participants expressed frustrations over having to operate within an environment where there was conflict between the various policies, processes and practice. Participants saw this conflict as arising from the inconsistencies in the 'myriad of legislation' as well as from the need to respond to a hierarchy of assessment bodies and/or organizations. The following quotes were typical of the sentiment in relation to complexity and over-regulation.

We're not observing simplification in the system, the complexity is increasing. (Proposer 2: architect)

That's just over-regulation on a massive scale. (Proposer 5: project manager)

High levels of regulation have implications for the design manager in various ways. For example, for a smaller residential housing subdivision or building project, the design manager may be the architect, the housing/land developer or project manager and depending on their role may need to know the details of many of these regulations. If the project is more substantial and the design team is larger, then the design manager would at least have to have enough knowledge of the various regulations to manage coordination of the negotiations and discussions.

We broke that down once to about 160 item issues that you should be considering in considering a development application ... there's our problem over the last few years ... it has been the myriad of legislation that keeps coming at us from all sorts of directions. (Proposer 1: architect)

This has implications for continuing professional education and development and tertiary education of undergraduates. It also has potential future implications for legal liability. In the short term, it has flow-on effects to each regulating authority individually requiring more documentation.

I have to say that getting the DAs [development approvals] is getting harder and harder ... it's getting to the point where they're asking for more and more ... more and more time consuming, more expensive – more and more documentation. (Proposer 6: development manager)

The objectives and requirements of specific individual authorities have legitimacy and the problems arise when one set of requirements from a specific authority diverge from the requirements of other authorities. There is obvious lack of agreement about what objective is of uppermost importance across the levels of government and coordination is problematic for design managers. Some specific examples include:

... this is the balance that we have to make that decision on, so you've got that legislation, you've got the bush fire legislation, you've got the vegetation legislation, the flora and fauna endangered species legislation and they all collide at some point in time. (Proposer 1: architect)

In essence, in a residential development we have four separate DCPs which are different. (Proposer 5: project manager)

I've been to state government, local government, I've been to both mayors ... he [the current mayor] doesn't want to talk to anybody, he knows everything there is about tanks ... But his chief adviser doesn't want to know about it because BASIX doesn't come under him and storm water's not his problem. (Proposer 1: architect)

BASIX – Building Sustainability Index – is a Web-based planning tool developed by the NSW Department of Planning (Robinson and Edwards, 2009). It rates the potential environmental performance of project developments using nine criteria including site, social amenity, transport, water, storm water, energy, waste and recycling, materials and indoor energy. It was originally developed in 2002 and is not dissimilar in aim to various other sustainability indices such as BREEAM (UK Building Research Establishment Environmental Assessment Method), LEEDS (US Leadership in Energy and Environmental Design) and BEPAC (Canadian Building Environment Performance Assessment Criteria).

Participants identified the inconsistencies in the prescriptive and non-prescriptive nature of the planning instruments across all cases as one of the key causes of conflicts. Despite this, the same participants who criticized the 'overly rigid' and prescriptive nature of the regulations also criticized the non-prescriptive nature of some regulations as 'lacking clarity'. Although an apparent contradiction on the part of these participants, this incongruous response reflects the deeper problems of lack of definition of sustainable urban development, lack of procedural guidelines for resolving conflicts and lack of categorization of issues on a scale of subjective to objective and corresponding prescriptive and non-prescriptive criteria and measures and criteria. Development applications often cannot be resolved to the satisfaction of all parties through the application of a single set of prescriptive 'rules' in planning instruments. The general perception was that aspects not clearly defined by the instruments, subject to the interpretation of those providing the approvals, were problematic. As highlighted by one participant:

> ... the Design Review Panel under the SEPP 65... Well I just think that's a disaster quite frankly ... the people that are seen on these panels have biased views ... as a professional in business we're all controlled by the DCPs, LEPs, REPs, SEPPs and they're the rules by which we've got to design and if we don't like them, well then we've got to make a move to try and have them changed in some way or another. But here you have this introduction of these design panels who have a pretty influential position on the outcome of an application ... my experience is that most of these people in the reports that come out of these Design Review Panels are in complete contradiction to the DCPs and the LEPs you know... (Proposer 1: architect)

Design review panels are a group of experts who provide independent specialist advice to assist local agencies on medium-density residential developments. Their advice is not binding but the reports have significant influence. Design schemes are presented during the early negotiation period prior to submitting a development application. This system was introduced just prior to the 2006 reforms to improve design quality.

This example of contradiction is a problem. Furthermore, the 'tactic' employed to resolve this contradiction would have a direct impact on the success of the development proposed. Ultimately, the two primary goals during the initial design management phase include, first, to achieve the stated objectives of the planning instruments and, second, to obtain the development approval within a reasonable timeframe. However, a design manager's ability to achieve these goals is made difficult when the planning advice is inconsistent, such as when the design review panels are 'in complete contradiction to the DCPs and LEPs'.

The capacity to produce a development application that satisfies both what is prescribed by the instruments as well as the subjectivity of those assessing the application is thus inevitably made more complex by the lack of clear definition by the instruments on what sustainable urban development means and on how to resolve the conflicts as they arise. Planners often provide conflicting advice as they attempt to 'second guess' politicians who will ultimately make the final decisions regarding approval. Planners involved in the study stated that they felt in many cases that their specialist advice to the council decision makers (i.e. democratically elected politicians) on proposals was ignored. As a result, design managers, who are largely driven by the need to minimize conflicts during the development approval process, find themselves frustrated in a reactive decision-making environment. They constantly negotiate, renegotiate and respond to the 'people that they're getting approval from' in order to convince them that the development proposed is indeed satisfying their requirements but all within an environment where the boundaries are shifting. This type of reactive behaviour encourages a climate of self-interest and persuasive opportunism.

There was general agreement among proposers and assessors that increased resources are required to monitor and respond to the various requirements associated with the development approval process. One proposer in particular expressed frustration in relation to the added time and costs associated with the volume of legislation which impacted negatively not only on the project costs and progress but also the design manager and client relationship. For this participant, the tension created in the design manager–client relationship was associated with a 'confidence thing' whereby the design manager's (in this case an architect) inability to obtain development approval in a timely manner reflected negatively on their credibility.

Consequently, design managers find themselves employing strategic tactics to expedite the approval process. Assessors, despite expressing the desire to 'work to regulations and consider every conceivable bit of legislation', found the reality of time and resource constraints resulted in 'certain times where ... [assessors had] ... to make a value judgement'. In such situations, assessors acknowledged that they may allow themselves to be persuaded by the various participants in the development approval system simply for the purposes of gaining a timely approval as opposed to arriving at the best solution in terms of sustainability. Some of the key persuasion tactics that emerged as participants respond and attempt to reduce the conflicts encountered throughout the development approval process are discussed in more detail in the following section.

PERSUASION TACTICS IN THE APPROVAL PROCESS

In summary, across all case studies the two most common persuasion tactics demonstrated in relation to the discourse on sustainability on projects were persuading through reciprocity and persuading through authority (Table 4).

Reciprocity was raised as a tactic commonly used by proposers and in particular the developers involved with the development approval process. This form of persuasion tactic involves the return of a favour by another who has previously been provided a favour. Within the development approval system, a past favour provided to the council by the developer

often in the form of donations is reciprocated by the council in the form of a vote in support of the developer's development application. As one stakeholder commented:

We know they [the council] have made their minds up of how they are going to vote because they have already caucused how they are going to vote, they have spoken to the developers in private meetings.

Although only raised by the community groups as a persuasion tactic used by the proposers, it is worth noting that it was a significant tactic, which was raised numerous times in the interviews in cases 1 and 3. There is a widely held perception of corruption by community stakeholders. In some localities, history has shown that the perception is reality. Corruption is the most extreme form of reciprocity; there are various levels of reciprocity. Even if conflicts of interest are made transparent either at the point of decision making or in public reports regarding declarations of conflicts of interest, in some localities any form of discussion between business community leaders and local politicians are viewed with a high level of suspicion. There will always be diverse interests in relation to proposals. There is a need for transparency and good governance as well as more rigorous ways of developing intellectual scientific arguments about sustainability merits of projects. This leads to the next persuasion tactic – persuading through authority.

Persuading through authority emerged as the most dominant theme across the three case studies. The analysis revealed a much richer way in which participants persuade through authority than that

TABLE 4 Key themes in relation to persuasion tactics described by the participants

THEMES	CASE 1	GROUP	CASE 2	GROUP	CASE 3	GROUP
Reciprocity	Favours	S			Favours	S
Authority	Quality Information	S P A	Quality Information	S P	Quality Information	A
	Expertise		Expertise	S	Disregard for expertise	P
	Position in community: financial, local action groups	S P	Position in community: local action groups	S	Position in community: financial, local action groups	P

S – Stakeholder; P – Proposer; A – Assessor

previously suggested to us by Cialdini (1993). Further to Cialdini's definition of authority in relation to persuasion we now include other key dimensions revealed, which are specific to the development approval system. The key dimensions identified include design information, professional expertise and position in a community.

The analysis highlighted the consistent use of information in the form of good quality documentation by proposers to persuade others in the development approval system. A proposer from case 2 in particular highlighted their competence and ability to develop 'first class documentation' as a general representation of the potential to achieve better quality developments and as an advantage for attracting clients. Indeed, the assessors interviewed expressed a preference for high-quality documentation and revealed a tendency towards being persuaded by the type of information provided through documentation whereby 'those that prepare very well can often persuade'. The potential value of a development is therefore judged on the basis of one's ability to persuade and 'prove that you've considered everything' as evidenced through the professionalism demonstrated in the development application documentation.

> I mean persuasion is about selling your message and yeah understanding what your issue is and I mean it's all in the preparation, some people prepare very well and some people don't and those that prepare very well can often persuade ... it's about putting your argument together and being persuasive but it's not about yelling or screaming ... it's about having those discussions if it works and if it doesn't work and being convincing about what should and shouldn't happen. (Proposer 3: architect)

Some other participants, however, particularly the stakeholders, challenged the credibility of these well-prepared applications produced by developers as 'so called artists' impressions as distinct from reality'. It was felt that the attractive presentation of development proposals through 'beautiful plans not drawn to scale' was simply a veneer employed to persuade others of the merit of the development and to ultimately obtain the necessary approvals.

Proposers also conceded to the argument underlying this observation whereby they admitted to the practice of 'second guessing how council's going to think'. Consultants would then include appropriately convincing information in the related documentation in order to minimize potential difficulties or conflicts in the development approval process. In doing so, it was recognized that the quality of the development may be compromised and the potential for introducing innovative and sustainable design solutions has been significantly restricted.

The second dimension of persuading through authority involves persuading through the individual specialization or skills of various professions including architects, land surveyors, developers, environmentalists, landscape architects, urban designers, water engineers, etc. The expertise of professionals was recognized as an effective persuasion tactic whereby an assessor described an example in which a major developer employed 'a very professional team of consultants' who got the development approval 'through a fairly quick process'. The assessor further commented that it was the appropriate skills or specialization demonstrated by the 'right sort of consultants' that was central to their ability to persuade. However, there was some disagreement between various research participants on this idea. While some are able to persuade successfully through expertise, others achieve less success when attempting to persuade despite having the appropriate skills. One participant voiced frustrations over the disregard for their architectural skills as a result of the assessing body being persuaded by other parties who are against the development proposed. For this proposer, it was felt that the development proposals that got approved were in no way an accurate representation of the value of a development where the 'merits of architecture just go to water' and it is simply a case of 'who has got the biggest lawyer'. This is one example where the use of one persuasion tactic appears to be more effective than another; however, this would require deeper investigation to validate.

The last effective authority persuasion tactic that emerged is persuading through one's position in a local 'community'. The community was varied and could include business, arts or lobbying groups. The position and power of local resident action groups has had a significant impact upon the persuasion environment in the development approval system. Their position within the local community is often used as a persuasion tactic to 'organize petitions, get signatures, hassle the council officers ... to cause a stir' against developments proposed and ultimately achieve what is perceived by them to be sustainable urban development. It is important to note that this tactic was described by many participants as one that is used fairly frequently and effectively by many local action groups to achieve their objectives in relation to sustainable urban development, which in this case was underpinned by an 'anti-development' approach.

There are other examples though and using one's position to influence decisions is not restricted to local resident groups. The financial position of major businesses was also seen to be useful when attempting to persuade. As commented by one participant, 'this whole conflict is money vs poverty if you like' indicating the capacity for such major companies with 'millions of dollars behind them' to lobby the associated bodies to achieve their objectives through the development proposed.

the bodies that actually implement these plans have changed ... it's always moving fast, it's pretty hard to keep in contact with. (Stakeholder 8)

The design manager has had to become particularly adept at working within an environment of persuasion. The 'persuasion movement' is the antithesis of the philosophy underpinning the 1980s collaborative design and participatory consultative approaches. During the 1980s, there was an attempt to involve all participants in the design and planning processes as a way towards creating environments that were more responsive towards the users' economic, social and environmental objectives. The persuasion environment has emerged from this early ideal of collaboration, to a situation where participants employ various tactics to ultimately pursue their own individual objectives rather than a balanced achievement of social, economic and environmental considerations. Often for the community stakeholder groups it is the social objectives that are at the forefront of their agenda,

whereas for the proposers it is typically the economic objectives that are driving their agenda. However, both the collaborative and persuasion *movements* are ultimately attempts to resolve conflicts which involve high levels of consultation and negotiation between the different participants.

IMPLICATIONS AND CONCLUSIONS FOR THEORY AND PRACTICE OF DESIGN MANAGEMENT ROLE

Design managers, seeking to minimize resources (costs and time) and streamline coordination and yet still achieve quality design throughout the development approval stages, attempt to respond to the requirements of authorities in order to reduce potential conflicts and ensure the project remains feasible. This focus on the reduction of conflicts can detract from the potential for improving aspects of a development project, such as design quality and amenity of urban areas. Indeed, this was raised by many of the proposers whereby it was felt that the need to comply with the strict guidelines prescribed by the various acts and regulations limited their capacity to introduce potentially innovative design solutions. Clearly everyone is impacted by negotiation and persuasion during the development approval process.

The central conflict of management of urban growth in an environmentally sensitive way and the achievement of local economic objectives through participatory and collaborative planning processes is well discussed in the literature on planning. The literature on planning as a participatory or a collaborative process seems overwhelming and diverse in perspectives and theoretically complex (Plöger, 2001). There are different approaches to managing the tension between economic and environmental development. Typically the extremes have been classified as the neo liberalist vs the managerialist/corporatist approach (Gleeson and Low, 2000). This is not only a philosophical divide between how to approach governance of development which gives rise to conflicts but also the quite pragmatic procedural and regulatory conflicts which occur.

Two interesting and divided theories of planning have emerged and are useful to us when theorizing

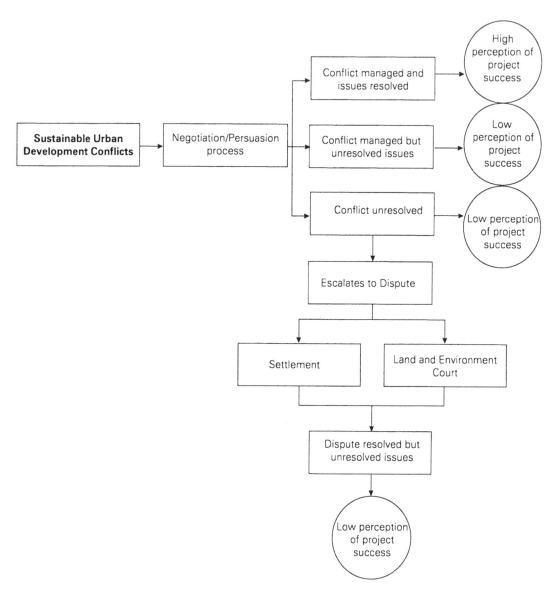

FIGURE 3 Conflict management in the development approval system (source: London and Chen, 2007)

about the practice of design management in relation to the planning phase (Plöger, 2001). First, theories of communicative and argumentative planning relying upon Habermasian philosophy of the ethics of participation and democratic communication (Habermas, 1984), i.e. a process of interactive collective reasoning (Healey, 1996). Second, the art of governance or the art of empowering urban community governance challenges communicative planning theorists and their idealized situation because urban governance favours technocratic knowledge and legal power, i.e. governance refers '... to the processes through which collective affairs are managed' (Healey, 1996). A Foucaultian critique of planning (Foucault, 1991) argues that it is not usually an inclusionary practice and it does not emphasize '... knowledgeable reasoning and argumentation'. Instead, according to Plöger (2001), planning

'... favours rule bound behaviour, hierarchical and structured bureaucracy, and emphasizes technical and legal reasoning based upon political objectives'.

Design managers do not develop planning instruments nor the governance structures but they are intrinsically a part of the negotiations involved in the *practice* of urban development. From a practitioners' perspective, tasked with integrating interdisciplinary information and knowledge, both professional and non-professional, there are important lessons to learn from both the Habermasian theory of communicative planning (Habermas, 1984) and the

Foucaultian art of governance discursive power relations approach (Foucault, 1991) and also this study and the empirical observations made.

In particular, an appreciation of the relationship between knowledge–power–participation is essential. Sustainable urban design is central to achieving sustainable urban development and design managers are in a position to improve a problematic development approval system. The following framework for design management is proposed in Figure 4 and is grounded in theory and empirical observations. There are four key parts to the framework including governance,

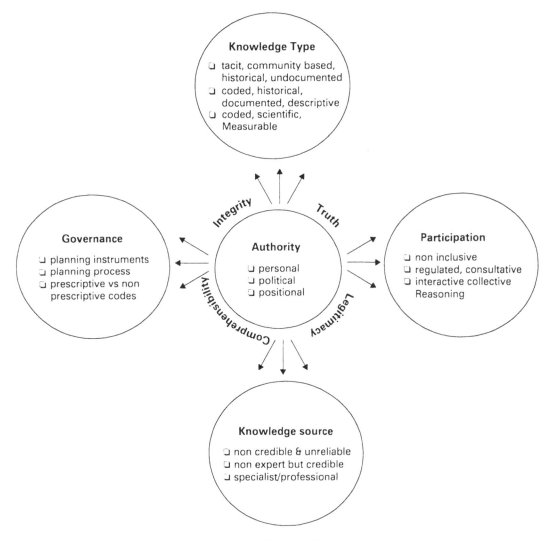

FIGURE 4 Inclusive knowledge-based design management for sustainable urban development

participation, knowledge type and knowledge sources. Each of these have attributes that can impact positively and/or negatively and which are more or less important in various developments. The aim of the framework is for design managers to act reflexively and to facilitate working towards developing argumentation for proposals that are based on the commonly accepted principles of judgements by Habermas (1984) including comprehensibility, integrity, legitimacy and truth of statements (Plöger, 2001) and respond to the multi-stakeholder voices.

Sustainable urban development is a contested ground and will always remain so. The design manager is increasingly required to take an inclusive approach. However, what does that really mean? This research has resulted in a more refined understanding of the implications of managing design through the early development approval phase. The challenge to the design manager is to address two key roles. First, the emergence of the role as a *by default* central informal *arbitrator* across agencies' conflicting planning instruments and second, as a *navigator* through a set of complex, varied and inter-related persuasion tactics used by all stakeholders. The design manager is typically working on behalf of a particular client and therefore is more than likely obliged to develop unique persuasion skills and capabilities in response to the individual characteristics of the project that they are involved in for their client's best interests. However, increasingly, there are changing societal ethical expectations of those involved in the property and construction industry in response to sustainable urban development. It is a capacity for self-reflection, awareness and ethical practices when negotiating and persuading that will mark the future design manager as society places greater expectations on those responsible for planning, designing and constructing our infrastructure.

AUTHOR CONTACT DETAILS

Kerry Anne London: Chair in Construction Management, Deakin University, Waterfront Campus, Geringhap St, Faculty of Science and Technology, School of Architecture and Building, Geelong VIC 3220, Australia. Tel: +61 3 5227 8358, fax: +61 3 5227 8341, e-mail: Kerry.London@deakin.edu.au

Katie Cadman: School of Architecture and Built Environment, Faculty of Engineering and Built Environment, The University of Newcastle, University Drive, Callaghan, NSW 2308, Australia

REFERENCES

Anderson, J., Nycyk, M., Jolley, L. and Radcliffe, D., 2005, 'Design management in a construction company', in D. Radcliffe and J. Humphries (eds), *Proceedings of the 2005 ASEE/AaeE 4th Global Colloquium on Engineering Education, Globalisation of Engineering Education Kindergarten to Year 12 Pipeline Transformation of the Disciplines, Australasian Association for Engineering Education, Sydney, Australia, September 2005*.

Australian Government Productivity Commission, 2004, *Reform of Building Regulation, Research Report*, www.pc.gov.au/__data/assets (accessed July 2008).

Bengston, D., Fletcher, J. and Nelson, K., 2003, 'Public policies for managing urban growth and protecting open space: Policy instruments and lessons learned in the United States', in *Landscape and Urban Planning*, 69, 271–286.

Cialdini, R., 1993, *Influence: The Psychology of Persuasion*, New York, Quill.

Development Assessment Forum, 1996, *Time For Business, Report of the Small Business Deregulation Task Force*, www.daf.gov.au/reports_documents/pdf/time_for_business.pdf (accessed July 2008).

Development Assessment Forum, 2005, *A Leading Practice Model for Development Assessment in Australia*, paper presented at the Development Assessment Forum, Canberra, ACT.

Development Assessment Forum, 2006, *Unfinished Business, Prospects for and Intergovernmental Agreement on Development Assessment*, www.daf.gov.au/reports_documents/pdf/unfinished_business.pdf (accessed February 2006).

Emmitt, S., 2007, *Design Management for Architects*, Oxford, Blackwell Publishing.

EDO (Environmental Defender's Office), 2008, *Submission on the Environmental Planning and Assessment Amendment Bill*, Sydney, Environmental Defender's Office.

Foucault, M. 1991, 'Governmentality', in G. Burchell *et al* (eds), *The Foucault Effect: Studies in Governmentality*, Chicago, IL, University of Chicago Press, 87–105.

Gleeson, B. and Low, N., 2000, *Australian Urban Planning: New Challenges, New Agendas*, Australia, Allen & Unwin.

Golobic, M. and Marusic, I., 2007, 'Developing an integrated approach for public participation: a case of land-use planning in Slovenia', in *Environment and Planning B: Planning and Design*, 34(6), 993–1010.

Gray, C.G. and Hughes, W., 2001, *Building Design Management*, Oxford, Butterworth-Heinemann.

Greene, R. and Elfrers, J., 1999, *Power the 48 Laws*, London, Profile Books.

Gurran, N., 2002, *Housing Policy and Sustainable Urban Development: Evaluating the Use of Local Housing Strategies in Queensland*, New South Wales, and Victoria Australian Housing and Urban Research Institute, University of Sydney Research Centre.

Habermas, J., 1984, *The Theory of Communicative Action, Volume 1*, Boston, MA, Beacon Press.

Healey, P., 1996, 'The communicative turn in planning theory and its implications for spatial strategy formations', in *Environment and Planning B: Planning and Design*, 23(2), 217–234.

Higgins, C., Judge, T. and Ferris, G., 2003, 'Influence tactics and work outcomes: a meta-analysis', in *Journal of Organisational Behaviour*, 24, 89–106.

Innes, J.E., 1993, 'Implementing state growth management in the United States: Strategies for coordination', in J.M. Stein (ed), *Growth Management: The Planning Challenge of the 1990s*, Newbury Park, CA, Sage, 18–43.

LGSA (Local Government and Shires Associations of NSW), 2008, *Local Government's Formal Response to Draft Exposure Bills (Planning Changes)* Local Government and Shires Associations of NSW, www.lgsa.org.au/www/html/2475-planning-changes.asp (accessed 20 October 2008).

London, K. and Chen, J., 2007, 'Conflict and persuasion in sustainable urban development decision making', in *Proceedings of the CIBW92 Building Across Borders Procurement Conference*, Hunter Valley, Australia, September 2007, 34.

Lyle, J.T., 1985, *Design for Human Ecosystems*, New York, Van Nostrand Reinhold.

Mayer, I., 1997, *Debating Technologies. A Methodological Contribution to the Design and Evaluation of Participatory Policy Analysis*, Tilburg, Tilburg University Press.

Mayer, I. and Seijdel, R., 2005, 'Collaborative decision making for sustainable urban renewal projects: a simulation ^ gaming approach', in *Environment and Planning B: Planning and Design*, 32(3), 403–423.

McKenzie, F., 1997, 'Growth management or encouragement? A critical review of land use policies affecting Australia's major exurban regions', in *Urban Policy and Research*, 15(2), 83–101.

Peters, B., 1998, 'Managing horizontal government: the politics of coordination', in *Public Administration*, 76(2), 295–311.

Plöger, J., 2001, 'Public participation and the art of governance', in *Environment and Planning B: Planning and Design*, 28(2), 219–241.

PCA (Property Council of Australia), 2006, *Reasons to be Fearful? Residential Development Costs Benchmarking Summary Report*, PCA Residential Development Council, www.affordablehome.com.au/files/ pdf/Reasons%20to%20be%20Fearful%20.pdf (accessed 4 March 2008).

PCA (Property Council of Australia), 2007, *Boulevard of Broken Dreams; The Future of Housing Affordability in Australia*, PCA Residential Development Council, www.affordablehome.com.au/files/pdf/Boulevardofbrokendreams.pdf (accessed 4 March 2008).

RAIA (Royal Australian Institute of Architects)/Archicentre, 2003, *Planning Assessment Survey Report: Report 2003*, www.archicentre.com.au/media/planningreport2003.pdf (accessed July 2008).

RAIA (Royal Australian Institute of Architects)/Archicentre, 2005a, *Planning Assessment Survey Report: Report 2005a: Report No 2*, www.architecture.com.au/i-cms?page=6791 (accessed July 2008).

RAIA (Royal Australian Institute of Architects)/Archicentre, 2005b, *Survey Confirms Australia's DA Processes Worsen*, Sydney, The Royal Australian Institute of Architects.

Robinson, D. and Edwards, D., 2009, 'Sustainable housing design: measurement, motivation, and management in Sutherland Shire, Sydney, Australia', in *Environment and Planning B: Planning and Design*, 36(2), 336–354.

Steiner, F., 1999, *The Living Landscape*, New York, McGraw-Hill.

Tewdwr-Jones, M. and Thomas, H., 1998, 'Collaborative action in local plan-making: planners' perceptions of "planning through debate"', in *Environment and Planning B: Planning and Design*, 25(1), 127–144.

Tibaijuka, A., 2002, 'Address by the Executive Director of UN-HABITAT to conference delegates', in *Sustaining our Communities*, papers from *International Local Agenda 21Conference*, Department for Environment and Heritage, South Australia, Adelaide, March 2002, www.regional.org.au/au/soc/ (accessed February 2009).

UTA (Urban Taskforce Australia), 2008, *A Planning System in Gridlock, How to Save NSW from a Social and Economic Crisis*, Sydney, Urban Taskforce Australia.

Watson, V., 2003, 'Conflicting rationalities: implications for planning theory and ethics', in *Planning Theory and Practice*, 4(4), 395–407.

Weitz, J., 1999, 'From quiet revolution to smart growth: State growth management programs, 1960 to 1999', in *Journal of Planning Literature*, 14(2), 267.

Williams, P., 2007, 'Planning and the legislative framework', in S. Thompson (ed), *Planning Australia: An Overview of Urban and Regional Planning*, Melbourne, Cambridge University Press.

Yukl, G., 1998, *Leadership in Organisations*, Sydney, Prentice-Hall.

ARTICLE

Public Procurement Incentives for Sustainable Design Services: Swedish Experiences

Josefin Sporrong and Jan Bröchner

Abstract

When procuring building design services, public clients may provide incentives for the development of design tools and management practices that lead to more sustainable buildings. Procedures for selecting design professionals, including the choice of contract award criteria, can be analysed as the outcome of external and internal factors that influence procuring units. This investigation studies how the selection procedures that Swedish municipal clients rely on when procuring services from architectural and engineering consultants provide incentives for environmentally sustainable design management. Questionnaire responses from 93 Swedish municipalities show that a minority include sustainability-related criteria in tender assessments. Environmental management systems were a criterion for 30% of the respondents, while only 11% regularly used life cycle cost as a criterion. More than half of all municipalities were reported to have a general policy for services procurement, and these policies usually include environmental sustainability. However, these policies are not always followed when procuring construction-related services. Smaller municipalities are less likely to have a general policy. Consequently, incentive effects for design service firms are probably weak. Better skills and provider-selection methods among clients are needed for rewarding sustainable design practices more strongly.

■ *Keywords* – Public procurement; design services; sustainability

INTRODUCTION

In its twin roles of regulator and buyer of construction and construction-related services, the public sector may create an institutional setting that provides incentives for designers to develop skills and practices that contribute to a more sustainable built environment. The inclusion of life cycle costing as a design tool, advocated since the 1960s (Stone, 1975) and currently more generally recognized (Gray and Hughes, 2001: 129ff; Kotaji *et al*, 2003; Tunstall, 2006: 287ff), and the introduction of an environmental management system in the design firm can be seen as examples of sustainability practices in design

management, preparing the way for more integrated approaches to sustainable design (Khalfan, 2006; Vakili-Ardebili and Boussabaine, 2007a).

Procurement is used increasingly as a political and corporate instrument to stimulate the environmental performance of products and services (Legarth, 2001; Marron, 2003; European Commission, 2004). Green procurement introduces environmental considerations into purchasing policies, programmes and actions (Russel, 1998). By expressing environmental preferences, purchasers can indirectly affect the environmental performance of products and services and thus reduce adverse impacts in an ecological sense (Verschoor and Reijnders, 1997; Zsidisin and Siferd, 2001). However, the procurement of construction design services is more of a challenge

ARCHITECTURAL ENGINEERING AND DESIGN MANAGEMENT ■ 2009 ■ VOLUME 5 ■ PAGES 24–35
doi:10.3763/aedm.2009.0903 © 2009 Earthscan   ISSN: 1745-2007 (print), 1752-7589 (online)

from Routledge

because design is one step removed from the actual products, i.e. the buildings, that will ultimately affect the environment. There is a recent case study of the relationship between public procurement of building design and a range of quality aspects (Volker *et al*, 2008), but sustainability in this context has not been in focus earlier. In the present investigation, 'sustainable design services' refers to design services that are part of or lead to sustainable construction (Hill and Bowen, 1997; Bourdeau, 1999) and we look primarily at ecological or environmental sustainability, sometimes loosely referred to as 'green'.

The aim of the investigation described here was to study how the selection procedures that Swedish municipal clients rely on when procuring services from architectural and engineering consultants provide incentives for environmentally sustainable design management. Based on findings from a questionnaire survey covering 93 municipalities, the analysis intends to give more insight regarding how environmental issues are perceived, framed and acted upon in the municipal procurement of building design services.

The article is structured as follows. First, a review of the literature on selection of service providers is presented, successively narrowed to earlier research on public procurement of construction-related services. The development of environmentally sustainability issues related to procurement in the Swedish construction sector is outlined. Next there is a description of the survey methodology. Key findings from the survey of municipal procurement officers are then reported. Conclusions are finally drawn, and a few policy recommendations are given.

SELECTING DESIGN SERVICE PROVIDERS

In principle, how design firms take environmental issues into account and let them shape their design management is likely to be influenced by contract award criteria that emphasize sustainability. The development and application of sustainability criteria in public procurement by local authorities can be assumed to reflect national policies, municipal policies and also the skills and attitudes of procurement officials. Thus the procedures for selecting design professionals, including the choice of contract award criteria, can be analysed as the outcome of a number of factors, both external and

internal, that influence procuring units in the public sector. Incentives are present on two levels: procurement officials are subject to incentives, weak or strong, that make them conform to national and local policies for sustainability; and officials create sustainability incentives for design firms by choosing them according to particular criteria. One external factor in the context of local procuring units is therefore the legal regulation of public procurement. Internal factors comprise procurement strategies and policies, including evaluation methods, as well as procurement skills. For public procurement officials, the internal factors are assumed to operate within the limits imposed by the legal framework.

THE LEGAL CONTEXT

Over the past decades, there has been a global trend towards increased legal formalization of public procurement to secure transparency, equal treatment and reduce barriers to cross-border trade in both goods and services. The wish to regulate the procurement also of advanced professional services and the need to recognize environmental consequences of patterns for public buying have led to the development of more complex award strategies involving multiple criteria rather than selecting providers according to lowest price. The OECD Greener Public Purchasing project (Marron, 2003) gives an international overview of how governments have begun including environmental criteria and changed their procurement routines in other ways.

Whereas the concept of sustainable development was included as an overarching objective in the 1997 EU Treaty, the inclusion of environmental requirements and criteria in public procurement appeared to need clarification. The 1992 (92/50/EEC) directive on procurement of public services did not specify the type of environmental requirements that could be included in public procurement, but the 2004 directive (EC/2004/18) indicates the possibilities of adopting environmental considerations in technical specifications, selection and award criteria as well as contract performance clauses (European Commission, 2004). Officials may award a contract either to the tender that is considered to be the economically most advantageous on the basis of multiple criteria or to the tender with the lowest price. If multiple criteria are to be used, there are rules to safeguard that they are applied transparently.

INTERNAL FACTORS: THE LOCAL PROCURING ENTITY

Studies of the purchasing function of companies and public authorities have indicated that organizational factors such as procurement strategies and skills are strongly associated with procurement performance.

Strategies and skills

Many authors have recognized a shift in the procurement function, away from its traditional administrative and transactional role to a more strategic role. Even if the extent of this shift has been disputed, it is now widely acknowledged that proactive value-focused rather than passive cost-focused procurement strategies are integral to long-term organizational strategy (Tassabehji and Moorhouse, 2008). However, researchers have shown more interest in the private sector in general and the manufacturing industry in particular, although the work of Murray (1999, 2001a, 2001b) forms an exception. Based on a survey of UK council leaders in 47 councils, the goals of local government, purchasing's potential strategic contribution and the strategies currently used were established (Murray, 2001a). His investigation showed that the discrepancy between the current and the potential contribution to the strategic goals of local government was significant. One key area of development that was identified was an increased contribution to the environment.

Jointly with this growing emphasis on strategic procurement, empirical research is emerging to demonstrate the impact of specific procurement skills on organizational performance. Thus skills have been described as having a direct influence on the ability of procurement professionals to fulfil their role proficiently (Tassabehji and Moorhouse, 2008). Giunipero and Pearcy (2000) highlight changes in the procurement function and the environment in which it operates as significantly affecting the ideal skills of procurement professionals. In the same vein, several authors have stressed the need for procurement staff to update existing skills and develop new perspectives and abilities in order to improve the organizational

contribution of the purchasing function (Henke, 2000; Giunipero et al, 2005, 2006; Cousins et al, 2006).

Other authors have studied differences between purchasing skills in the private and the public sectors. The more extensive scrutiny of public procurers' activities (Gordon et al, 2000; Mechling, 1995) makes knowledge of national and European regulations an important procurement competence. Public scrutiny tends to keep public purchasers focused on how people perceive their adherence to procedure, rather than on whether value for money spent has been achieved (Pettijohn and Qiao, 2000). In addition, public buyers typically have a more limited experience of procurement of consultancy services in general (Corcoran and McLean, 1998; Smeltzer and Ogden, 2002; Roodhooft and Van den Abbeele, 2006).

Design and construction procurement criteria

Much of what has been written on procurement of construction and of related design services has centred on the strategic issue of criteria used for selecting tenders. In traditional public sector procurement, the award of contracts is based merely on a comparison of tender prices (Palaneeswaran et al, 2003). This is also a common method of choice for the procuring of construction design services in the public sector (Christodoulou et al, 2004). Similarly, in a study carried out by Pottinger (1998), representatives from both the private and the public sectors clearly viewed public sector clients as more focused on price than private sector clients. Several researchers (Palaneeswaran et al, 2003; Christodoulou et al, 2004) have proposed alternative selection procedures for the award of construction-related contracts, procedures that include the assessment of both price and non-price criteria (which may include environmental criteria). The obvious logic behind this is that low bid selection does not guarantee the overall lowest project cost upon project completion (Wong et al, 2001), nor the lowest cost during the life cycle of a building (Christodoulou et al, 2004). In addition, a focus on lowest price for construction design services may not provide best quality or the highest satisfaction among clients (Ling, 2004). But there is also a broader range of internal factors to be considered: the effectiveness of using environmental requirements in the procurement of road maintenance has been related to a number of factors such as a committed management, a well functioning organization with a broad competence and a clearly identified policy towards green purchasing (Faith-Ell *et al*, 2006).

THE CASE OF SWEDEN

Both Swedish public procurement and environmental awareness went through considerable change during a few years in the early 1990s, coinciding with the European integration of Sweden. As a member state of the European Union, Sweden has implemented the directives on public procurement in the Public Procurement Act, although the legal changes due to the 2004 directive came into force only in 2008.

Two major initiatives to promote sustainable construction in Sweden have been taken by the Ecocycle Commission, appointed by the Swedish government in 1993, and the Ecocycle Council for the Building Sector, established in 1994, which includes developers, property owners, architects and consultants to the building industry. The Ecocycle Council has ranked significant environmental impacts from buildings, based on life cycle assessment analyses (Byggsektorns Kretsloppsråd, 2001). The Royal Swedish Academy of Engineering Sciences (IVA, 1997) advocated that construction clients should be a strong force in improving technologies for sustainability, and central government began using public procurement strategically. Organizations developed goals and policies for environmental management, such as implementing environmental management systems (EMS) and including environmental specifications in their procurement processes (Stenberg and Räisänen, 2006). Thus a government–industry dialogue initiative was set up in 1998 and a working group has published recommendations for procurement in a life cycle perspective (Building, Living and Property Management for the Future, 2003).

A set of 15 (today 16) national environmental quality objectives was launched in Government Bill 1997/98:145 and adopted by the Swedish Parliament in 1999. One of these objectives is for a good built environment, but a recent assessment by the Environmental Objectives Council (2008) indicates that this ambitiously formulated objective is very difficult or impossible to reach in the time stipulated.

Partly, the good built environment objective addresses design management: 'Buildings and amenities must be located and designed in accordance with sound environmental principles and in such a way as to promote sustainable management of land, water and other resources'.

A survey undertaken by the Swedish Environmental Protection Agency reveals the current status of environmental procurement efforts among public authorities (Naturvårdsverket, 2008). Of all municipalities, 82% have environmental procurement guidelines, but only 62% actually follow these guidelines. The survey showed that 55% of the municipal procurers had undergone special training in environmental procurement. According to this investigation, the main obstruction for environmental procurement, as perceived by all public authorities, was lack of knowledge and experience of how to formulate environmental requirements. Other frequently mentioned obstacles were the additional costs arising when including environmental considerations in procurement procedures, that it is time consuming, the perception of the public procurement regulations as complex, as well as lack of interest within the organization.

Sterner (2002) performed an early investigation of which environmental aspects Swedish private and public clients considered when procuring buildings. She emphasized that environmental requirements used in the building sector necessitate that the purchasing organization develops environmental strategies, such as policies and guidelines, which in turn are reflected in purchasing practices, such as the adoption of environmentally conscious evaluation methods. In another study, interviews with 29 procurement officers in eight municipalities during the years 2003–2005 were carried out to investigate the selection of construction contractors (Carlsson and Waara, 2006). Procurement officers were found to prefer environmental criteria that are easy to evaluate, such as whether a bidder has a corporate environmental policy or an environmental management system. Barriers to integrating environmental concerns were identified as lack of administrative resources, including environmental know-how, as well as restricted budgets and uncertainty regarding the legislation, resulting in risk-averse behaviour among procurement officers.

A survey of 386 construction contracts, procured by local authorities in 2003, showed that multi-criteria selection, rather than lowest price, was used in more than 80% of the cases (Waara and Bröchner, 2005, 2006). In contrast to an earlier study of how municipalities select providers of architectural services (Lindqvist, 2001), this implies that local authorities have started to award contracts to contractors, architects and consultants based more on 'value' than low fees. For central government authorities, and especially in the case of professional construction-related services, this is a frequent practice; in case studies carried out by Sporrong et al (2005), the objective was to investigate how large public clients procured architectural and engineering services, particularly how they used other criteria than price when awarding contracts, and interviews with these clients showed a clear preference for multi-criteria selection methods for professional services. Furthermore, the implementation of environmental requirements in government road maintenance contracts has been studied by Faith-Ell et al (2006). A systematic approach including both the procurement and the implementation phases of a contract was seen as necessary for efficiency in applying these requirements. Environmental indicators and clearly stated environmental requirements were found to be important in combination with follow-up procedures.

Against this background, there are four main research questions that concern incentives for sustainable design management. First, is there a local procurement policy, and does it include elements of sustainability? Second, which sustainability criteria for selecting providers of architectural and engineering services are used in practice? The third question to be answered is the nature and the strength of the link between a general procurement policy in the municipality and its procurement practice for architectural and engineering services. Finally, the fourth question concerns perceived needs for improved local practices.

METHODOLOGY

For the survey described here, procurement officials were identified in about half of all 290 municipalities in Sweden, chosen randomly. In Swedish municipalities, the procurement of construction-related services is typically performed jointly by a purchasing department, which usually has a central and more purely administrative function, and a technical services department. Alternatively, procurement of construction-related services is found in only one of these entities. Most officials in the goal population belong to a purchasing or a technical services department, being managers and experts responsible for the procurement of architectural and engineering services. However, some municipalities have outsourced the purchases of construction-related services to a company owned by the municipality. There are also some municipalities, usually small, who have chosen to share a common purchasing function.

A six-page postal questionnaire was sent to the officials in 2007. The survey instrument, an explorative questionnaire, was divided into five sections: respondent background, purchasing organization of the client, tender evaluation procedures, purchasing policy and potential areas of development. Together with multiple choice questions, five-degree Likert scales were used to measure respondent opinions of a number of questions and statements. Local application of contract award criteria that refer to environmental management systems and life cycle cost are examples of what we have considered to indicate concern with sustainable design management. A first version of the questionnaire was piloted with three persons in two municipalities not included in the random sample. These persons contributed to the final version primarily by reacting to the structure of the survey and the phrasing of questions. Ultimately, and after reminders, answers had been received from 93 out of 130 questionnaires distributed, equivalent to a response rate of 72%. The responses cover 37% of the Swedish population, which implies that the size distribution of municipalities in the sample is representative.

SURVEY FINDINGS

Respondents belonged to organizations of different sizes; the largest municipality included in the sample had almost 300,000 inhabitants, the smallest one barely 3000. This spread in population figures can be expected to be mirrored in procurement strategies and skills. Experience of procurement varied: of all respondents, 46% had been involved in an average of 10 to 59 selection processes for architectural and engineering services. However, as many as one in three respondents had handled a maximum of only nine processes, which may reflect that many small municipalities are included in the sample. One in five respondents were more experienced, having participated in at least 60 selection processes.

LOCAL PROCUREMENT POLICIES

The questionnaire section dealing with purchasing policy included: (a) whether there exists a municipal procurement policy; (b) the content of the policy; and (c) to what extent the stipulated requirements in this policy are adhered to, according to the respondent. Responses were given on a five-degree Likert scale ranging from 'never' (= 1) to 'always' (= 5).

As reported by 56% of all respondents, their municipality currently had a general procurement policy for services. According to 43%, environmental considerations were included in the policy. One third of respondents stated that this policy was complied with at least to some degree (answers spread over 'sometimes', 'often' and 'always'). Sustainability was stated by one respondent in five as being part of the policy, and 16% indicated that this policy requirement was followed more or less in practice (again referring to answers 'sometimes', 'often' and 'always'). Results show that the environmental and sustainability policy requirements are affirmatively adhered to 'often' or 'always' by only 25% and 12% of the respondents, respectively.

Further, the results show that 41% of the municipalities have a procurement policy that includes total economy (basically to be understood as life cycle cost), while 31% of the respondents stated that this policy is more or less followed in practice, with a stronger emphasis on 'often' and 'always'. Moreover, 24% have included long-term thinking in their procurement policy. According to 17% of the respondents, their municipality also complies with this policy item, with a slightly stronger emphasis on 'often' and 'always'.

Results for municipalities grouped according to population are displayed in Table 1. In order to avoid problems of how respondents interpret the word 'sustainability' (in Swedish, hållbarhet) in the questionnaire, a range of related concepts were listed: life cycle cost, long-term issues, environment in addition to sustainability. Clearly, the smaller

TABLE 1 Municipal service procurement policies and their application to design contract award, by municipal population (percentage for each population range, or average frequency on a five-degree scale, where 1 = never applied, 5 = always applied)

ASPECT	POPULATION 50,000 AND ABOVE	POPULATION 10,000–49,999	POPULATION 2000–9999
No. of responding municipalities	19	58	16
Municipality has a services procurement policy	74%	52%	50%
If there is a policy:			
Policy includes life cycle cost	79%	80%	37%
Frequency of applying life cycle cost policy to procurement of design	4.67	4.59	3.75
Policy includes long term issues	64%	40%	13%
Frequency of applying long term policy to procurement of design	4.60	3.89	3.33
Policy includes environment	93%	77%	50%
Frequency of applying environmental policy to procurement of design	4.44	4.19	4.20
Policy includes sustainability	57%	37%	25%
Frequency of applying environmental policy to procurement of design	4.17	4.00	3.67

municipalities rely less on general policies for procurement, and when there is a policy that includes environmental aspects, they appear to be less enthusiastic than larger municipalities about applying it.

CHOICE OF SELECTION CRITERIA

Corresponding to the second main research question, there were survey questions regarding the general set of selection criteria for providers of architectural and engineering services in the municipality. The results show that the existence of an EMS is a criterion for provider choice in 30% of the municipalities. However, only 11% of the municipal respondents stated that they use a life cycle cost (LCC) criterion on a regular basis when selecting architects and engineers.

Furthermore, the respondents were asked to estimate (on a five-degree Likert scale) the difficulty of applying their general selection of criteria in practice. Interestingly, 36% of the respondents perceive the assessment of a potential provider's EMS as difficult or somewhat difficult (31% find it both easy and difficult; 4% find it difficult and 1% find it very difficult). For the LCC criterion, the proportion of those who think this criterion is difficult or somewhat difficult to evaluate is somewhat lower (27%). The number of those who think this criterion is difficult to evaluate is higher however (8%) than for the EMS criterion, as well as those who think it is very difficult (3%).

Another question included statements regarding choice of selection criteria in general (Table 2). The statements: 'Although evaluation is according to "economically most advantageous", price is often decisive' and 'In general, we attach too much weight to price' were supported by as many as 78% and 61%, respectively.

In this context, the figures fail to indicate that the size of the municipality is correlated with opinions.

INDIVIDUAL ATTITUDES

Is there a link between general policy documents for procurement in a particular municipality and its reliance on LCC criteria in actual procurement practice? This is a more precise version of the third main research question. The following analysis is based on procurement procedures for architectural services only.

Eleven respondents stated that they normally rely on an LCC criterion for procuring architectural services. In Tables 3 and 4, differences between municipalities that do apply an LCC criterion and those who do not are shown. Although there is a link between general policy documents for procurement that include sustainability and long-term thinking in a particular municipality and its reliance on LCC criteria in actual procurement practice, the link is weak (Table 3), especially if we consider how a general goal of 'sustainability' fails to be reflected in the choice of criteria. Policy statements that emphasize 'total economy' in procurement appear to be much more easily translated into a practice that evaluates LCC capabilities among architects.

Of those procurement officials who consider that procurement of architectural services functions within their own municipality less than 'very well' (5) or 'well' (4) on a five-degree Likert scale, those who apply an LCC criterion are less prone (average 2.83) to

TABLE 2 Perceptions of price effects and of central government sustainability policies, by municipal population (averages on five-degree scale, where 1 = disagree fully and 5 = agree fully)

STATEMENT	POPULATION 50,000 AND ABOVE	POPULATION 10,000–49,999	POPULATION 2000–9999
'Although evaluation is according to "economically most advantageous", price is often decisive'	3.68	3.55	3.53
'We often rely on lowest price because we wish to avoid judicial reviews'	2.50	2.33	2.71
'In general, we attach too much weight to price'	3.11	2.94	3.38
'The public sector does too little for creating a sustainable Sweden'	2.56	2.76	2.50

TABLE 3 Municipalities applying an LCC criterion to architectural procurement compared with other municipalities

GENERAL MUNICIPAL PROCUREMENT POLICY INCLUDES	APPLY LCC CRITERION	DO NOT APPLY LCC CRITERION
Total economy	64%	40%
Long-term thinking	36%	22%
Sustainability	27%	24%

TABLE 4 Views of procurement officials in municipalities applying an LCC criterion to architectural procurement and of officials in other municipalities (averages on a five-degree scale, where 1 = do not agree, 5 = agree fully)

STATEMENT	APPLY LCC CRITERION	DO NOT APPLY LCC CRITERION
'The public sector does too little for creating a sustainable Sweden'	3.09	2.64
'Too many publicly procured buildings suffer from faults'	3.00	2.72
'It should be possible to find a better method for measuring creativity of architectural services'	3.91	3.65

agree with a statement that there is a lack of competence for assessing architectural services. Staff from municipalities who do not rely on an LCC criterion show an average of 3.29 on this five-degree scale. In addition, it should be noted that it is clearly more frequent among procurement officials in municipalities using an LCC criterion to have personal experience of working as an architect or consultant.

It seems that the existence of a general procurement policy that prioritizes sustainability and long-term thinking in a municipality provides only a partial explanation for local criteria in practical use. An alternative explanation, which does not exclude influence from a common policy, is that local procurement officials hold individual opinions that are mirrored in their procedures for procurement. In Table 4, differences in how officials agree to three relevant statements can be seen. Among municipalities that do not apply an LCC criterion, the

average reaction of officials is rather that the public sector slightly exaggerates the issue of sustainability, which is in contrast to the almost neutral or slightly affirmative response from those who have put LCC thinking into procurement practice. There is also more of a complacent attitude towards faults in buildings among the non-LCC officials, although we cannot be sure that this does not result from a history of fewer faults in these municipalities; the introduction of an LCC criterion locally could be the consequence of bad experiences with earlier building projects. How satisfied are procurement officials with current methods for recognizing architectural creativity? It turns out that officials who do not apply an LCC criterion view the possibilities to develop better methods for the measurement of creativity as somewhat lower than their colleagues in LCC applying municipalities do.

AREAS OF DEVELOPMENT: METHODS AND SKILLS

Finally, the survey included questions regarding potential areas for development in the procurement of architectural and engineering services, answered by about 30% of the respondents who had stated that procurement of these services did not work fully satisfactorily. More than one in three (36%) of the respondents believe that existing procurement competence for the assessment of architectural services within the municipality is inadequate (on the five-degree scale, spread among the three options 'I partially agree'; 'I almost fully agree' and 'I fully agree'). For engineering services, the corresponding figure is only slightly lower (34%). As possible consequences of deficiencies in their current procurement practices (out of which procurement competence was given as one option among several others), 38% of the respondents stated 'uncertainty among assessors how tenders regarding architectural and engineering services should be evaluated' and 28% stated 'an excessive focus on price'.

DISCUSSION

Although the recent investigation by the Swedish Environmental Protection Agency (Naturvårdsverket, 2008) shows that no less than 82% of the municipalities have environmental procurement guidelines, we have found that only 43% of the municipalities have included environmental considerations in their general procurement policies for services. This discrepancy is odd, since our response rate is high and additionally there may be a non-response bias among municipalities that lack strong environmental policies. In addition, according to our survey responses, only one third of the municipalities comply, more or less, with their own general policies for including environmental considerations in procurement. One interpretation is that procurement officials are immune to general procurement policies, be they environmental or other. In other words, we have to distinguish between sustainability in procurement policy and sustainability in the application of procurement policy.

In their actual procurement practice for design services, the existence of an environmental management system was a criterion for 30% of the municipalities, and life cycle costs were taken into consideration by 11%. It is thus possible to speak of public procurement as an eco-design driver, which does not contradict the observation that underlying issues such as energy efficiency are strong drivers (Vakili-Ardebili and Boussabaine, 2007b).

General skills in purchasing professional services are important together with sustainability know-how, as identified by earlier investigations (Carlsson and Waara, 2006; Naturvårdsverket, 2008). Responses to the present survey showed that evaluation of environmental parameters is perceived as more or less difficult by one third of the local procurement officials, demonstrating that individual competence related to environmental assessment varies. A general lack of construction-related procurement competence is further underlined by the fact that more than a third of the respondents believe that existing competence for the assessment of architectural and engineering services within their municipality is inadequate. About as many respondents saw 'uncertainty among assessors how tenders for architectural and engineering services should be evaluated' and 'an excessive focus on price' as possible consequences of skill deficiencies. The finding that the emphasis on price is thought to be excessive among municipalities in general, a statement that was supported by as many as 78% of

the respondents, indicates that improved skills in assessing non-price criteria, such as those related to sustainable design, is needed among municipal procurers. In this way, the results of this investigation confirm earlier studies pointing at the public sector in particular as needing improved purchasing skills (Corcoran and McLean, 1998; Smeltzer and Ogden, 2002; Roodhooft and Van den Abbeele, 2006).

CONCLUSIONS

Our survey results indicate that many municipal clients in Sweden include sustainability criteria in their evaluation of architects and engineers, but far from all municipalities. The results also suggest that individual attitudes among municipal procurers could have an effect on procurement routines. Given the procurement officers' opinions recorded here, it seems that the existence of an environmental procurement policy offers only a partial explanation for requirements and selection criteria in practical use. Here is something that should be explored more in detail.

So which are the incentive effects on design management? Only a few environmental selection criteria have been found here to be in common use for evaluating potential architectural and engineering service providers. It is difficult to claim that the incentive effect for design service providers is strong, and there are probably other forces that are stronger in encouraging providers to introduce sustainable design management practices. The investigation presented here shows that there is a need for a higher awareness of how environmental sustainability policies can generate sustainable design management in firms that deliver construction-related services to local government. Our survey findings suggest that there is scope for many municipalities to learn from those who have introduced more advanced methods for assessing service providers so that sustainable practices in design management are rewarded more obviously. One remedy for an existing lack of relevant procurement skills among purchasing officials is to let them undergo training to familiarize them with their responsibilities for implementing environmental sustainability policies in their procurement procedures. The varying adherence to existing policies among municipal buying entities implies a need for the development of a common practice for applying sustainability policy requirements. In particular, smaller municipalities need more support for their development and application of procurement strategies that encourage design services providers to improve their sustainability skills.

Since the issue of sustainable building design and green procurement has been high on the Swedish agenda since the early 1990s, the present study of current practices should be of wider international interest. However, many of the clauses found in the European directives have been applied in the Swedish Act also to service contract sums well below the directive threshold values, and such calls are not advertised in the *Official Journal* (OJEU). Furthermore, when analysing how Swedish local authorities develop their procedures, it has to be kept in mind that central government is and has been reluctant to guide and monitor local practices, in procurement and in many other policy fields. Obviously, it is not just a Swedish phenomenon that there is a link between public client strategies and sustainable design management; studies of architects and design firms in Singapore give similar indications (Ofori *et al*, 2000; Ofori and Kien, 2004).

In summary, this investigation shows that in order to advance our understanding of sustainable design management, it is necessary to explore more deeply the interplay between environmental agendas, regulatory directives, organizational policies and strategies, as well as individual skills and opinions of important buyers of design services, not least in the public sector, and how these factors influence the realization of environmentally sustainable building projects. The incentive effect of sustainability policy driven procurement on the design management strategies of architects and technical consultants is one of several starting points for future research in this area.

ACKNOWLEDGEMENTS

Support for this investigation from the Swedish Research Council for Environment, Agricultural Sciences and Spatial Planning (Formas), the Chalmers Centre for Management of the Built Environment, the ARQ Foundation for Architectural Research, the J. Gust. Richert Memorial Fund and STD, the Swedish Federation of Consulting Engineers and Architects, is gratefully acknowledged. Helpful comments have been made by Lina Carlsson and Fredrik Waara, as well as by the anonymous referees.

AUTHOR CONTACT DETAILS

Josefin Sporrong: Department of Technology Management and Economics, Chalmers University of Technology, SE-412 96 Göteborg, Sweden. Tel: +46 76 248 85 57, e-mail: josefin.sporrong@chalmers.se

Jan Bröchner (corresponding author): Department of Technology Management and Economics, Chalmers University of Technology, SE-412 96 Göteborg, Sweden. Tel: +46 31 772 54 92, e-mail: jan.brochner@chalmers.se

REFERENCES

Bourdeau, L., 1999, 'Sustainable development and the future of construction: a comparison of visions from various countries', in *Building Research and Information*, 27(6), 354–366.

Building, Living and Property Management for the Future, 2003, *System Selection and Procurement with a Life Cycle Perspective*, Stockholm, Naturvårdsverket.

Byggsektorns Kretsloppsråd, 2001, *Byggsektorns betydande miljöaspekter: miljöutredning för byggsektorn*, www.kretsloppsradet.com (accessed 28 July 2008).

Carlsson, L. and Waara, F., 2006, 'Environmental concerns in Swedish local government procurement', in G. Piga and K.V. Thai (eds), *Advancing Public Procurement: Practices, Innovation and Knowledge-Sharing*, Boca Raton, FL, PrAcademics Press, 239–256.

Christodoulou, S., Griffis, F.J., Barrett, L. and Okungbowa, M., 2004, 'Qualifications-based selection of professional A/E services', in *Journal of Management in Engineering*, 20(2), 34–41.

Corcoran, J. and McLean, F., 1998, 'The selection of management consultants: how are governments dealing with this difficult decision? An explorative study', in *International Journal of Public Sector Management*, 11(1), 37–54.

Cousins, P.D., Lawson, B. and Squire, B., 2006, 'An empirical taxonomy of purchasing functions', in *International Journal of Operations and Production Management*, 26(7), 775–794.

Environmental Objectives Council, 2008, *Sweden's Environmental Objectives: No Time to Lose*, www.miljomal.nu/english/english.php (accessed 28 July 2008).

European Commission, 2004, *Buying Green! A Handbook on Environmental Public Procurement*, Luxembourg, Office for Official Publications of the European Communities.

Faith-Ell, C., Balfors, B. and Folkeson, L., 2006, 'The application of environmental requirements in Swedish road maintenance contracts', in *Journal of Cleaner Production*, 14(2), 163–171.

Giunipero, L.C. and Pearcy, D.H., 2000, 'World class purchasing skills: An empirical investigation', in *Journal of Supply Chain Management*, 36(4), 4–13.

Giunipero, L.C., Denslow, D. and Eltantawy, R., 2005, 'Purchasing/supply chain management flexibility: Moving to an entrepreneurial skill set', in *Industrial Marketing Management*, 34(6), 602–613.

Giunipero, L.C., Handfield, R.B. and Eltantawy, R., 2006, 'Supply management's evolution: Key skill sets for the supply manager of the future', in *International Journal of Operations and Production Management*, 26(7), 822–844.

Gordon, S.B., Zemansky, S.D. and Sekwat, A., 2000, 'The public purchasing profession revisited', in *Journal of Public Budgeting, Accounting and Financial Management*, 12(2), 248–71.

Gray, C. and Hughes, W., 2001, *Building Design Management*, Oxford, Butterworth Heinemann.

Henke, J.W., 2000, 'Strategic selling in the age of modules and systems', in *Industrial Marketing Management*, 29(3), 271–284.

Hill, R.C. and Bowen, P.A., 1997, 'Sustainable construction principles and a framework for attainment', in *Construction Management and Economics*, 15(3), 223–239.

IVA, 1997, *Byggherren i fokus*, Stockholm, Ingenjörsvetenskapsakademien.

Khalfan, M.M.A., 2006, 'Managing sustainability within construction projects', in *Journal of Environmental Assessment Policy and Management*, 8(1), 41–60.

Kotaji, S., Schuurmans, A. and Edwards, S., 2003, *Life-Cycle Assessment in Building and Construction: A State-of-the-art Report, 2003*, Brussels, SETAC.

Kumaraswamy, M.M. and Dissanayaka, S.M., 2001, 'Developing a decision support system for building project procurement', in *Building and Environment*, 36(3), 337–349.

Legarth, J.B., 2001, 'Internet assisted environmental purchasing', in *Corporate Environmental Strategy*, 8(3), 269–274.

Lindqvist, T., 2001, *Kvalitet eller pris? Upphandling av arkitekttjänster enligt Lagen om offentlig upphandling (LOU)*, Stockholm, Arkus.

Ling, F.Y.Y., 2004, 'Consultancy fees: Dichotomy between A/E's need to maximize profit and employers' need to minimize cost', in *Journal of Professional Issues in Engineering Education and Practice*, 130(2), 120–123.

Marron, D., 2003, 'Greener public purchasing as an environmental policy instrument', in *OECD Journal on Budgeting*, 3(4), 71–105.

Mechling, J., 1995, *Information Technology and Government Procurement: Priorities for Reform*, John F. Kennedy School of Government, Harvard University, Cambridge, MA.

Murray, J.G., 1999, 'Local government demands more from purchasing', in *European Journal of Purchasing and Supply Management*, 5(1), 33–42.

Murray, J.G., 2001a, 'Local government and private sector purchasing strategy: a comparative study', in *European Journal of Purchasing and Supply Management*, 7(2), 91–100.

Murray, J.G., 2001b, 'Improving purchasing's contribution: The purchasing strategy of buying council', in *International Journal of Public Sector Management*, 14(5), 391–410.

Naturvårdsverket, 2008, *Tar den offentliga sektorn miljöhänsyn vid upphandling? En enkätstudie 2007*, Stockholm, Naturvårdsverket.

Ofori, G., Briffett, C., Gang, G. and Ranasinghe, M., 2000, 'Impact of ISO 14000 on construction enterprises in Singapore', in *Construction Management and Economics*, 18(8), 935–947.

Ofori, G. and Kien, H.L., 2004, 'Translating Singapore architects' environmental awareness into decision making', in *Building Research and Information*, 32(1), 27–37.

Palaneeswaran, E., Kumaraswamy, M. and Ng, T., 2003, 'Targeting optimum value in public sector projects through 'best value'-focused contractor selection', in *Engineering, Construction and Architectural Management*, 10(6), 418–431.

Pettijohn, C. and Qiao, Y., 2000, 'Procuring technology: issues faced by public sector organisations', in *Journal of Public Budgeting, Accounting and Financial Management*, 12(1), 441–61.

Pottinger, G., 1998, 'Property services: the private sector response to competitive tendering', in *Property Management*, 16(2), 92–102.

Roodhooft, F. and Van den Abbeele, A., 2006, 'Public procurement of consulting services', in *International Journal of Public Sector Management*, 19(5), 490–512.

Russel, T., 1998, 'Introduction', in T. Russel (ed), *Greener Purchasing: Opportunities and Innovations*, Sheffield, Greenleaf Publishing, 9–19.

Smeltzer, L.R. and Ogden, J., 2002, 'Purchasing professionals' perceived differences between purchasing materials and purchasing services', in *Journal of Supply Chain Management*, 38(1), 54–70.

Sporrong, J., Bröchner, J. and Kadefors, A., 2005, *Anbudsvärdering vid offentlig upphandling av arkitekt-och byggkonsulttjänster – förstudie*, CMB och Avdelningen för Service Management, Chalmers Tekniska Högskola, Göteborg.

Stenberg, A.-C. and Räisänen, C., 2006, 'The social construction of 'green building' in the Swedish context', in *Journal of Environmental Policy and Planning*, 8(1), 67–85.

Sterner, E., 2002, 'Green procurement of buildings: a study of Swedish clients' considerations', in *Construction Management and Economics*, 20(1), 21–30.

Stone, P.A., 1975, *Building Design Evaluation: Costs-in-Use*, 2nd edn, London, E&FN Spon.

Tassabehji, R. and Moorhouse, A., 2008, 'The changing role of procurement: Developing professional effectiveness', in *Journal of Purchasing and Supply Management*, 14(1), 55–68.

Tunstall, G., 2006, *Managing the Building Design Process*, 2nd edn, Amsterdam, Butterworth-Heinemann.

Vakili-Ardebili, A. and Boussabaine, A.H., 2007a, 'Creating value through sustainable building design', in *Architectural Engineering and Design Management*, 3(2), 83–92.

Vakili-Ardebili, A. and Boussabaine, A.H., 2007b, 'Design eco-drivers', in *Journal of Architecture*, 12(3), 315–332.

Verschoor, A.H. and Reijnders, L., 1997, 'How the purchasing department can contribute to toxics reduction', in *Journal of Cleaner Production*, 5(3), 187–191.

Volker, L., Lauche, K., Heintz, J.L. and de Jonge, H., 2008, 'Deciding about design quality: design perception during a European tendering procedure', in *Design Studies*, 29(4), 387–409.

Waara, F. and Bröchner, J., 2005, 'Multicriteria contractor selection in practice', in K. Sullivan and D. Kashiwagi (eds), *CIB W92 International Symposium on Procurement Systems, Las Vegas, Tempe, AZ, PBSRG – Arizona State University, February 2005*, 167–172.

Waara, F. and Bröchner, J., 2006, 'Price and non-price criteria for contractor selection', in *Journal of Construction Engineering and Management*, 132(8), 797–804.

Wong, C.H., Holt, G.D. and Harris, P., 2001, 'Multi-criteria selection or lowest price? Investigation of UK construction clients' tender evaluation preferences', in *Engineering, Construction and Architectural Management*, 8(4), 257–271.

Zsidisin, G.A. and Siferd, S.P., 2001, 'Environmental purchasing: A framework for theory development', in *European Journal of Purchasing and Supply Management*, 7(1), 61–73.

ARTICLE

Corporate Social Responsibility of Architectural Design Firms Towards a Sustainable Built Environment in South Africa

Ayman Ahmed Ezzat Othman

Abstract

The construction industry makes a vital contribution to the social and economic development of every country. Buildings provide their users with places for housing, education, culture, medication, business, leisure and entertainment. None of these buildings will perform its function unless supported with efficient road networks, superlative telecommunications facilities, water and electricity. On the other hand, the construction industry has major impacts on the environment. It is a very large consumer of non-renewable resources, a substantial source of waste, pollution, land dereliction and energy consumption. This highlights the responsibility of present generations to use the available resources in a way that enables them to meet their needs without compromising the ability of future generations to meet their own needs. Since architects are one of the main players in the construction industry, this research aims to investigate the corporate social responsibility (CSR) of South African architectural design firms (SAADF). This was achieved through a questionnaire and a small number of interviews with respondents. Combined, this provided a unique insight into an important aspect of sustainable design management.

■ *Keywords* – Corporate social responsibility; architectural design firms; sustainability; built environment; South Africa

INTRODUCTION

The construction industry, in terms of its activities and output, is an integral part of every country's economic growth and social development process. Economically, it contributes towards achieving national goals and basic needs as well as providing most of the country's fixed capital assets and infrastructure that enable other industrial sectors to develop. In addition, it helps to increase a country's gross domestic product (GDP), thereby stimulating further growth via its linkages with other industrial sectors and creating job opportunities (Field and Ofori, 1988; Mthalane *et al*, 2007). Socially, the construction industry plays a pivotal role through

constructing projects that provide society with places for housing, education, culture, medication, business, leisure, entertainment as well as urban infrastructure such as water and power supply, sewerage, drainage, roads, ports, railways and telecommunications (Friends of the Earth, 1995; Roodman and Lenssen, 1995; Khan, 2008).

Construction also has a major impact on the environment. It is estimated that 3 billion tonnes of raw materials and 40% of the total flow into the global economy are used in manufacturing construction materials throughout the world (Roodman and Lenssen, 1995). The construction sector is responsible for 50% of the material resources taken from nature, 40% of energy consumption and 50% of total waste generated. Energy is an invisible resource that is consumed during the procurement of materials

ARCHITECTURAL ENGINEERING AND DESIGN MANAGEMENT ■ 2009 ■ VOLUME 5 ■ PAGES 36–45
doi:10.3763/aedm.2009.0904 © 2009 Earthscan   ISSN: 1745-2007 (print), 1752-7589 (online)

and construction activities. The former accounts for between 70 and 90% of the energy consumed in construction (Anink et al, 1996). Large amounts of materials and energy are also wasted in constructing and operating artificial heating and cooling systems.

Current generations have the right to use available resources to achieve their goals and meet their expectations. However, using these resources inefficiently compromises the ability of future generations to meet their own needs. As the first line of contact with clients in the construction industry, and being responsible for designing buildings and specifying construction materials (Othman, 2008), architects have an important role to play in corporate social responsibility (CSR). The work reported in this paper investigated the CSR of South African architectural design firms (SAADF) and provides a unique insight into an important aspect of sustainable design management.

THE CORPORATE SOCIAL RESPONSIBILITY CONCEPT

Charity and philanthropy are not new ideas (Rockey, 2004), although the field of CSR has grown considerably over the past decades and many businesses have become more active in contributing to society. CSR issues are now being integrated into all aspects of business operations and explicit commitment to CSR is made in the visions, missions and value statements of an increasing number of companies all over the world. CSR reports usually go beyond profit maximization to include the company's responsibilities to a broad range of stakeholders including employees, customers, community and the environment (Ofori and Hinson, 2007). In his speech in March 2006, Malcolm Wicks, the UK's energy minister with responsibility for CSR, stated that 'successful companies will be the ones that continually seek to raise their game and take a responsible approach to all their activities. These activities contribute to a kind of triple bottom line: ecological, financial and social' (Freshfields Bruckhaus Deringer, 2006).

A significant number of terms and definitions are used, including; corporate responsibility, corporate accountability, corporate ethics, corporate citizenship, sustainability, stewardship, triple bottom line and responsible business (Hopkins, 2004). CSR could be defined as the voluntary integration of environmental, social and human rights considerations into business operations, over and above legal requirements and contractual obligations (Freshfields Bruckhaus Deringer, 2006). CSR is the commitment of an organization to act in a manner that serves the interests of its stakeholders (Schermerhorn et al, 2005) and is concerned with the ways that companies generate profits and their impact on the broader community (Bradshaw and Vogel, 1981). It is about how companies manage their business processes to produce an overall positive impact on society (Baker, 2007). McAlister (2005) and Carroll (1993) mentioned that there is a widespread acceptance of the view that if a business is to prosper, then the environment in which it operates must prosper as well. This means that business must adopt approaches in which companies see themselves as part of a wider social system.

Over the years, CSR has been developed from the classical 'profit-centred model' to the modern 'socially responsible model' (see Carroll (1999) for a comprehensive overview). The classical model states that the management's only legitimate goal is to maximize profit. Milton Friedman (1962, cited in Ofori and Hinson, 2007), who has been recognized as an advocate of this view, believes that the primary responsibility of managers and directors is to operate in the best interests of the shareholders who are essentially the true owners of a corporation. The classical view perceives that corporate expenditure on social causes is a violation of management's responsibility to shareholders at least to the extent that these expenditures do not lead to higher shareholder wealth. On the other hand, Frank Abrams (1954, cited in Ofori and Hinson, 2007) stated that a firm's management is responsible for maintaining an equitable and working balance among the claims of the various directly interested groups such as stockholders, employees, customers and the public at large.

The World Business Council for Sustainable Development (WBCSD, 1999) viewed CSR as a continuous commitment by business to behave ethically and contribute to economic development while improving the quality of life of the workforce, their families and society at large. CSR means different things to different stakeholders, and Baker (2007) highlighted that in different countries there will be different priorities and values that will shape how business undertakes its CSR. The WBCSD in its publication *Making Good Business Sense* also highlighted some evidence of the different perceptions of CSR from a number of different societies across the world. Recently, Othman and Mia (2008) investigated the practical application of CSR and developed an innovative framework that integrated CSR into the quantity surveying profession as an approach to support the government initiatives for housing the poor in South Africa.

SUSTAINABILITY

The existence of more than 70 different definitions for sustainability (Pearce *et al*, 1989; Holmberg and Sandbrook, 1992) highlighted its importance and showed the efforts made by different academic and practical disciplines to define and understand its implications to their fields. Nevertheless, all definitions agree that it is essential to consider the future of the planet and find creative ways to protect and enhance the Earth while satisfying various stakeholders' needs (Boyko *et al*, 2006). The most commonly used definition is derived from the Brundtland Commission, which defined sustainability as development that meets the needs of the present without compromising the ability of future generations to meet their own needs (WCED, 1987). This people-centred definition focused on three main quality-of-life objectives. These have become known as the three dimensions of sustainability (social, environmental and economic), comprising:

● social progress that addresses the need for all people
● the effective protection of the environment and prudent use of natural resources
● the maintenance of stable levels of economic growth and development (DETR, 2000).

Sustainable architecture aims to reduce the negative environmental impacts of the constructed buildings throughout the project life cycle. It is about using the architect's talent and technical knowledge to design and build in harmony with the environment. The challenge is about finding the balance between environmental considerations, society requirements and economic constraints (Hui, 2002).

Sustainable design, and sustainable design management, aims to deliver sustainable architecture. It is the art of designing physical objects that comply with the principles of environment, society and economy. It is a growing trend within the fields of architecture, landscape architecture, engineering, industrial and interior design. Sustainable design aims to produce products and services in a way that reduces the use of non-renewable resources, minimizes environmental impact and relates people with the natural environment (Levin, 1995). Sustainable design is a general reaction to the global environmental crisis, i.e. rapid growth of economic activity and human population, depletion of natural resources, damage to ecosystems and loss of biodiversity (Shu-Yang *et al*, 2004).

There are three principles of sustainable design, namely:

● economy of resources
● life cycle design
● humane design.

Each principle embodies a unique set of strategies that can help architects perceive architecture's interaction with the environment and hence enable specific methods to be developed that could be applied to reduce the environmental impact of our built environment.

ARCHITECTURAL DESIGN FIRMS, CSR AND SUSTAINABILITY

In spite of their important role as the first line of contact with clients in the construction industry and being responsible for designing buildings and specifying construction materials, architects have a duty not only to their clients, but also to society and the environment at large. Architects play a significant role towards producing buildings and facilities that save the environment, enhance society and prosper the economy. Hence, their role should not be confined to the technical aspects, but to cover other aspects that extend their role to improve the sustainability of the built environment. In order to

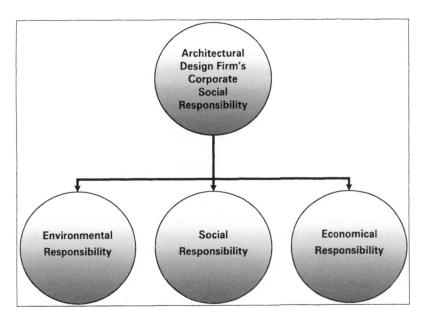

FIGURE 1 The CSR of architectural design firms

relate the CSR of design firms to sustainability, the social role of design firms will be discussed from three perspectives (Figure 1).

ARCHITECTURAL DESIGN FIRMS AND THE ENVIRONMENT

The construction process represents a major contributor to climate change, resources depletion, pollution and energy consumption at both local and global levels (Ofori et al, 2000; Addis and Talbot, 2001), hence, integrating environmental requirements into the design decision-making process is essential for producing buildings that save the environment. The CSR of architectural design firms towards the environment can be identified as:

- Escalating the awareness of the importance of saving the environment and encouraging the adoption and application of sustainability concepts in the architecture practice.
- Reducing the negative environmental impact of buildings through using durable, environment-friendly, non-toxic, easy to maintain, energy efficient and recyclable construction materials and equipment.

- Responding positively to the different environmental effects, forces and unexpected events (e.g. earthquakes, floods, climate constraints) through developing designs that cope with these constraints.
- Encouraging design firms and other construction professionals to be proactive in taking their responsibilities within their supply chains.

ARCHITECTURAL DESIGN FIRMS AND THE SOCIETY

Architecture is an integral part of human activities. It affects everyday experiences and actions. Architects are facing a challenge that is characterized by creating an environment that supports, enhances and celebrates human activities. Cities, towns and buildings have always been the result of cultural, social and economic factors. This requires the architect to be responsible towards these factors which support the design of a responsive environment. The CSR of architectural design firms towards society can be identified as:

- Raising the awareness of the important role that design firms can perform towards improving the society.

- Perceiving stakeholders' requirements and involving them in the design decision-making process to ensure that the developed facilities meet their needs, fulfil their expectations and reduce the cost and implications of later modifications.
- Equipping buildings with facilities for people with special needs as well as health and safety requirements.
- Seeking feedback from people who are affected by the built environment, providing support and adding value to communities and the supply chain.
- Including CSR in the architectural education programmes and providing expert advice to non-experts through offering volunteer services.
- Promoting positive partnerships between the public sector and SAADF to support government initiatives, improve collaboration and experience exchanges.
- Offering training programmes and jobs for recently graduated architects and engineers as well as sponsoring students.

ARCHITECTURAL DESIGN FIRMS AND THE ECONOMY

The economic dimension of sustainable architecture can be seen from two perspectives. First, stimulating growth in the construction industry which increases the percentage of GDP and provides more job opportunities. Second, increasing clients' profit and investment returns. The CSR of design firms towards the economy can be identified as:

- Highlighting the importance of the role that architectural design firms can play to improve the economy.
- Ensuring that society's funds and resources are used sparingly.
- Promoting and supporting purchasing from supply chains that are committed to sustainability requirements in their products.
- Considering the life cycle cost of the project and minimizing the cost of operation and maintenance.
- Creating innovative ideas and using sustainable materials and technology which can perform the same function or even better at lower cost.

- Specifying locally manufactured materials to encourage the national economy and reduce the cost of importing materials.
- Using demolition materials in manufacturing new construction materials.
- Creating job opportunities to reduce the unemployment rate and enhance families' economic status.

RESEARCH METHODS

A field study comprising a survey questionnaire and interviews was carried out on a sample of South African architectural design firms. A total population of 1354 firms, which are registered as members of the South African Institute of Architects (SAIA, 2008), were identified. These firms are distributed in 11 branches throughout South Africa, as shown in Table 1. A systematic random sample using a sample factor of (1:10) was applied to get a reasonable sample size and ensure that the study sample is distributed in the same proportion as the population in terms of the different branches (Bernard, 2000; Bryman, 2001). This resulted in producing 137 units to be surveyed. The survey questionnaire was then faxed and e-mailed to these firms. The sample size suits the population taking into account a 95% confidence level and 7.94 sampling error (Creative Research Systems, 2008). Design firms that responded to the survey questionnaire were then invited to participate in interviews.

Two approaches were used for data analysis. First, the quantitative approach adopted the measure of central tendency and dispersion to analyse the questionnaires and interview responses. The measure of central tendency was used to get an overview of the typical value for each variable by calculating the mean, median and mode. The measure of dispersion was used to assess the homogenous or heterogeneous nature of the collected data by calculating the variance and the standard deviation (Bernard, 2000). Second, since there is no quantification without qualification and no statistical analysis without interpretation (Bauer and Gaskell, 2000), during the course of this research qualitative data analysis was employed. Analysis of the collected data showed close values of means, medians and modes, indicated typical central values and showed also low values of variance and standard

TABLE 1 Population and sample size of surveyed institutes of architects in South Africa

INSTITUTE BRANCH	NO. OF REGISTERED FIRMS	SAMPLE SIZE (1:10)
Border Kei Institute of Architects (BKIA)	46	5
Cape Institute for Architecture (CIA)	412	41
Eastern Cape Institute of Architects (ECIA)	46	5
Free State Institute of Architects (FIA)	56	6
Gauteng Institute for Architecture (GIFA)	302	30
KwaZulu-Natal Institute of Architecture (KZNIA)	203	20
Mpumalanga Institute of Architects (MPIA)	29	3
Northern Cape Institute of Architects (NCAI)	9	1
Limpopo Institute of Architects (LIA)	26	3
Northwest Province Institute of Architects (NWPIA)	10	1
Pretoria Institute of Architecture (PIA)	215	22
Total	1354	137

deviation. This confirmed the quality and the homogeneity of the collected data as well as the reliability of the research findings.

RESULTS

137 questionnaires were sent to architectural design firms throughout South Africa, of which 45 were completed and returned, providing a response rate of 33%. The response rate is typical for a survey questionnaire, lying within the 30–40% range which is usually deemed acceptable because so few people respond to questionnaires (Fellows and Liu, 1997). Although 45 is a small sample it does give a unique insight into the perceptions of architects in South Africa towards CSR and sustainability.

QUESTIONNAIRE RESPONSES

The questionnaire was divided into three sections. The first section collected general information about the firm. It included the organization name, contact details and designation of the respondents. All respondents were either heads of architectural departments, senior architects or architects. This ensured the consistency and relevance of the received information. The second section was about CSR. It focused on investigating the perception and application of CSR as well as the internal governance of design firms. In addition, it was designed to identify the reasons that encourage/hinder design firms from adopting CSR and the form of their CSR practices (if any). Furthermore, on a scale of 5, design

firms were asked to rank their awareness of CSR and to identify the use of an ethics officer. The third section was concerned with sustainability. It covered the understanding of the concept of sustainability and the awareness of SAADF towards developing sustainable built environment in South Africa. Findings of the survey questionnaires showed that:

- 34 out of 45 respondents to the survey questionnaire stated that they recognize the CSR concept and implement it in performing their daily business functioning. 50% of these firms mentioned that CSR is reflected in their companies' visions for the future, and the remaining firms stated that it is reflected in their mission statements.
- With reference to the reasons that encourage SAADF to adopt CSR, 75% of respondents stated that CSR is the right thing to do, and the remaining 25% of respondents added that CSR is a marketing tool that could be used to promote design firms.
- Figure 2 shows the form of CSR practices carried out by SAADF. The firms that mentioned 'others' stated that their community liaison officer works with local schools on a number of activities and their firms set targets for employing homeless people.
- Respondents who indicated that their firms do not adopt CSR attributed that to the lack of knowledge of the concept and understanding of

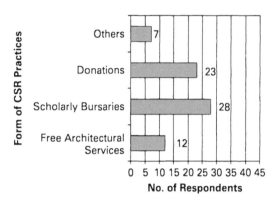

FIGURE 2 Responses of forms of CSR practices carried out by SAADF

their social role as well as the time, money and energy constraints incurred when adopting CSR.

● On a scale of 5, 18 out of 45 respondents to the survey questionnaire rated their awareness of CSR as 5 out of 5. 22 design firms rated their awareness as 4 out of 5, and the remaining firms rated their awareness as 3 out of 5.

● In order to investigate the use of ethics officers in design firms, 10 respondents out of 45 stated that they appoint an ethics officer as a liaison between the firm and the community.

● All SAADF indicated that they perceive and use the concept of sustainability in their business. 44.44 per cent of respondents rated the importance of their role towards developing a sustainable built environment as 5 out of 5, 53.33% rated their role as 4 out of 5, while the rest rated their role as 3 out of 5.

FINDINGS OF THE INTERVIEWS

Respondents were asked to attend a follow-up interview. Out of 45 design firms invited, five agreed to participate. Analysis of the interview responses showed that limited government incentives for supporting CSR as well as the partial evidence of successful sustainable projects hindered the adoption and application of CSR and sustainability in South Africa. In addition, interviewees stated that beyond the traditional role of SAADF, they can play a social role towards developing a sustainable built environment throughout the project life cycle as explained below.

During the feasibility stage

● Educating clients and explaining the importance of sustainability and its benefits to their projects, stakeholders and community in general.

● Integrating the different dimensions of sustainability during the briefing stage through understanding the client business case and value system and locating them in a sustainable context.

During the pre-construction period

● Incorporating sustainability requirements in the design process through critically analysing design elements and eliminating any item that does not add value to the project and may affect the final product and surrounding environment.

● Simplifying design to ease construction, reduce cost and facilitate maintenance.

● Specifying construction materials and equipment that comply and adhere to sustainability requirements.

● Providing equal opportunity and transparency in the selection process of contractors through comparisons between contractors based on their expertise and records of delivering sustainable products rather than lowest price.

● Obligating contractors to use sustainable materials and support suppliers that deal with such materials as well as automating the tender process and reducing paper work.

During the construction and after practical completion period

● Assisting contractors through finding creative solutions and alternative construction methods.

● Reducing energy consumption and wastage as well as applying health and safety regulations.

● Conducting post-occupancy evaluation to assess building performance and identify design errors and construction defects.

● Recording stakeholders' feedback for improving performance in future projects.

In addition, design firms identified a number of social roles that they can play in governmental, legal and technological spheres towards developing a sustainable built environment in South Africa. More specifically:

In the governmental sphere
● Creating partnerships with government authorities and supporting initiatives for housing the poor in South Africa through developing sustainable, affordable and quality housing projects.
● Explaining the benefits of sustainability to decision makers to get ownership of the concept and facilitate its adoption.
● Assisting government authorities in reviewing its regulations periodically to make sure that they are in line with sustainability requirements (Abdellatif and Othman, 2006).

In the legal sphere
● Proposing mandatory laws for implementing sustainability requirements, by all construction professionals, in the built environment.

In the technological sphere
● Ensuring that technology does not increase the unemployment rate.
● Employing new advanced and successful technology in building design to enhance the performance and reduce the consumption of materials and energy.
● Utilizing the benefits of information management and information technology to facilitate the procurement process and improve communications between concerned parties.

CONCLUSIONS AND THE WAY FORWARD
The development of CSR over the years and the increasing demand for integrating the environmental, social and economic considerations into the design decision-making process called for activating the social role of design firms to create innovative and value-added solutions towards developing a sustainable built environment in South Africa. In the research reported above, it was found that the concept of CSR is perceived and applied by most of the respondents, although different forms of CSR

practices are reportedly being implemented. The obstacles that hindered some design firms from adopting CSR and appointing ethics officers are attributed to a number of reasons such as lack of integrating CSR in the internal governance of design firms, limited government incentives for supporting CSR and the negative perception of time, cost and energy constraints when CSR is adopted.

The way forward to overcome these barriers and activate the CSR of SAADF towards a sustainable built environment in South Africa could be achieved by addressing the issues on two levels, namely the government and the design firm.

AT THE GOVERNMENT LEVEL
● Raising the awareness of CSR and educating the public and private sectors of the importance of CSR, as well as explaining its benefits towards developing a sustainable built environment in South Africa, are essential. This needs to be integrated with the establishment of appropriate rules and regulations that save the environment and encourage current generations to use natural resources sparingly to enable future generations to achieve their goals and meet their expectations.
● Supporting the private sector and creating partnerships with the SAADF to utilize their expertise and CSR towards developing sustainable projects could also be a positive development.

AT THE ARCHITECTURAL DESIGN FIRM LEVEL
Design managers clearly have a role to play in implementing the appropriate policies and providing the support and encouragement to carry through the initiatives. Some of those initiatives could include:

● Highlighting the importance of CSR and sustainability within the architecture profession and playing a proactive role beyond donations and scholarly bursaries and free architectural services.
● Integrating CSR in the internal governance of design firms and developing action plans and appropriate tools for implementing and

measuring its success as well as corrective actions required.

- Appointing ethics officers to manage the implementation of CSR in design firms and liaise between them and the community at large.
- Publicizing success stories and providing solid evidence that sustainability can be achieved within the client's budget and avoiding treating sustainability as a discrete problem.
- Integrating the concept of sustainability at the different stages of the project life cycle and particularly at early stages to obtain optimum advantages.

Although the research was limited to a small number of responses in comparison with the total population of architectural design firms in South Africa, the findings provide a valuable insight into how architects perceive and implement CSR towards achieving a sustainable built environment. More research is needed into this emergent aspect of design management to better understand the motives and drivers behind the issues revealed by this research.

ACKNOWLEDGEMENT

The author is grateful to the anonymous referees and the editorial board for their valuable suggestions.

AUTHOR CONTACT DETAILS

Ayman Ahmed Ezzat Othman: Senior Lecturer, School of Civil Engineering, Surveying and Construction, Faculty of Engineering, University of KwaZulu-Natal, King George V Avenue, Durban 4041, South Africa. Tel: +27 (0)31 2602821, fax: +27 (0)31 2601411, e-mail: Othman@ukzn.ac.za or aaeothman@gmail.com

REFERENCES

Abdellatif, M.A. and Othman, A.A.E., 2006, 'Improving the sustainability of low-income housing projects: The case of residential buildings in Musaffah Commercial City in Abu Dhabi', in *Emirates Journal for Engineering Research*, 11(2), 47–58.

Addis, B. and Talbot, R., 2001, *Sustainable Construction Procurement: A Guide to Delivering Environmentally Responsible Projects*, CIRIA C571, London, CIRIA.

Anink, D., Boodtra, C. and Mark, J., 1996, *Handbook of Sustainable Development*, London, James and James.

Baker, M., 2007, *Corporate Social Responsibility – What Does it Mean?*, www.mallenbaker.net/csr/CSRfiles/definition.html (accessed 3 February 2007).

Bauer, M. and Gaskell, G., 2000, *Qualitative Researching with Text, Image and Sound: A Practical Handbook*, London, Sage Publications.

Bernard, H.R., 2000, *Social Research Methods: Qualitative and Quantitative Approaches*, London, Sage Publications.

Boyko, C.T., Cooper, R., Davey, C.L. and Wootton, A.B., 2006, 'Addressing sustainability early in the urban design process', in *International Journal of Management of Environmental Quality*, 17(6), 689–706.

Bradshaw, D. and Vogel, T., 1981, *Corporations and Their Critics: Issues and Answers to the Problems of Corporate Social Responsibility*, New York, McGraw-Hill.

Bryman, A., 2001, *Social Research Method*, London, Oxford University Press.

Carroll, A.B., 1993, *Business and Society: Ethics and Stakeholder Management*, Cincinnati, OH, Southwestern Publishing Co.

Carroll, A.B., 1999, 'Corporate social responsibility: Evolution of a definitional construct', in *Business Society*, 38, 268–295, http://bas.sagepub.com/cgi/content/abstract/38/3/268 (accessed 3 February 2007).

Creative Research Systems, 2008, *Sample Size Calculator*, www.surveysystem.com/sscalc.htm (accessed 14 October 2008).

DETR (Department of Environment, Transport and the Regions), 2000, *Building a Better Quality of Life: A Strategy for More Sustainable Construction*, London, DETR.

Fellows, R. and Liu, A., 1997, *Research Methods for Construction*, London, Blackwell.

Field, B. and Ofori, G., 1988, 'Construction and economic development – a case study', in *Third World Planning Review*, 10(1), 41–50.

Freshfields Bruckhaus Deringer, 2006, *The Development and Impact of CSR on the Construction Industry*, www.freshfields.com/publications/pdfs/2006/16830.pdf (accessed 12 August 2008).

Friends of the Earth, 1995, *Prescription for Change: Health and the Environment*, Brussels, Friend of the Earth.

Holmberg, J. and Sandbrook, R., 1992, 'Sustainable development: What is to be done?', in J. Holmberg (ed), *Policies for a Small Planet*, London, Earthscan.

Hopkins, M., 2004, *Corporate Social Responsibility: An Issues Paper*, www.ilo.org/public/english/bureau/integration/download/publicat/4_3_285_wcsdwp-27.pdf (accessed 31 January 2007).

Hui, S.C.M., 2002, *Sustainable Architecture*, www.arch.hku.hk/research/BEER/sustain.htm#1.3 (accessed 11 April 2007).

Khan, R.A., 2008, 'Role of construction sector in economic growth: Empirical evidence from Pakistan economy', in *Proceedings of the First International Conference on Construction in Developing Countries (ICCIDC), Karachi, Pakistan, August 2008*, www.neduet.edu.pk/ICCIDC-I/Conference%20Proceedings/Papers/030.pdf (accessed 14 October 2008).

Levin, H., 1995, *Building Ecology: An Architect's Perspective on Healthy Buildings*, http://buildingecology.com/article.php (accessed 15 October 2008).

McAlister, D.T., 2005, *Business and Society: a Strategic Approach to Social Responsibility*, Boston, US, Houghton Mifflin.

Mthalane, D., Othman, A.A.E. and Pearl, R.G., 2007, 'The economic and social impacts of site accidents on the South African society', in J.J.P. Verster and H.J. Marx (eds), in *Proceedings of the 5th Post Graduate Conference on Construction Industry Development, Bloemfontein, South Africa, March 2008*, 1–10.

Ofori, D. and Hinson, R., 2007, 'Corporate social responsibility (CSR) perspectives of leading firms in Ghana', in *Corporate Governance*, 7(2), 178–193.

Ofori, G., Briffett, C., Gang, G. and Ranasinghe, M., 2000, 'Impact of ISO 14000 on construction enterprises in Singapore', in *Construction Management and Economics*, 18(8), 935–947.

Othman, A.A.E., 2008, 'Building the effective architectural team in design firms: The case of the United Arab Emirates', in *Emirates Journal for Engineering Research*, 13(1), 1–11.

Othman, A.A.E. and Mia, B., 2008, 'Corporate social responsibility for solving the housing problem for the poor in South Africa', in *Journal of Engineering, Design and Technology*, 6(3), 237–257.

Pearce, D., Markandya, A. and Barbier, E.B., 1989, *Blueprint for a Green Economy*, London, Earthscan.

Rockey, N., 2004, 'Breaking new ground in corporate social investment', in B. Bowes and S. Pennington (eds), *The Story of Our Future South Africa 2014*, Johannesburg, The Good News (Pty), Johannesburgh, South Africa.

Roodman, D.M. and Lenssen, N., 1995, *A Building Revolution: How Ecology Health Concerns are Transforming Construction*, Paper 124, Washington, DC, World Watch Institute.

SAIA (South African Institute of Architects), 2008, www.saia.org.za/ (accessed 14 October 2008).

Schermerhorn, J.R., Hunt, J.G. and Osborn, R.N., 2005, *Organizational Behaviour*, 9th edn, John Wiley and Sons, Hoboken, New Jersey.

Shu-Yang, F., Freedman, B. and Cote, R., 2004, 'Principles and practice of ecological design', in *Environmental Reviews*, 12(2), 97–112.

WBCSD (World Business Council for Sustainable Development), 1999, *Corporate Social Responsibility*, Geneva, WBCSD.

WCED (World Commission on Environment and Development), 1987, *Our Common Future*, Oxford, Oxford University Press.

ARTICLE

Promotion of Materials and Products with Sustainable Credentials

Matthew Peat

Abstract

The use of materials and products with sustainable credentials could potentially improve resource productivity and aid sustainable development. In order for such items to be widely adopted, the construction industry would need to overcome the barriers common to specification practice. The conservative nature of specifiers has been described by previous research, whereby familiar items are recurrently selected from a trusted palette and alternative solutions are only sought when the palette fails. The body of literature recognizes manufacturers as influential sources of information and has also shown that specifiers often make informal contact with companies known to them when faced with a specification problem. However, some existing work indicates that the active selection phase is not the only time when the decisions of specifiers can be influenced. Promotional material encountered during a state of passive awareness could be retained for later use, thereby expanding the specifiers' palette. This paper reports and discusses the results of a quantitative inquiry into the extent of publicity currently afforded to materials and products with sustainable credentials in mainstream construction industry periodicals. The findings reveal that problems in the marketing of these items may be acting as a barrier to their uptake by restricting the specifiers' awareness of the available options.

■ *Keywords* – Building product marketing; specification; sustainable materials; sustainable products

INTRODUCTION

The UK government (Performance and Innovation Unit, 2001) has identified the need to improve resource productivity as fundamental to achieving sustainable development. The objective is to increase economic activity while limiting the depletion of natural resources. It has been acknowledged (Performance and Innovation Unit, 2001) that the construction industry has a negative effect on the situation. The premise of this paper is that the use of materials, components and products with sustainable credentials could enable the construction industry to make a positive contribution to the government's agenda. These items encompass those directly reused, reclaimed, recycled, down-cycled or obtained from renewable sources.

Existing research (Chick and Micklethwaite, 2004) indicates that the use of 'green' items is not common

practice and suggests a number of obstacles to their adoption. These barriers include the lack of information relating to the types of materials and products available, the availability and location of suppliers and technical knowledge required for successful integration. It could be argued that the selection and specification of 'green' items will also be subject to the same issues prevalent in general specification practice. Therefore, a deeper insight into the barriers preventing the use of green items must be sought if those managing the detailed design stage of building projects are to influence greater resource productivity. This aligns with the suggestion by Emmitt (2006) that design managers must gain a sound appreciation of specification practices during the detailed design stage in order to be effective in managing the process.

ISSUES IN SPECIFICATION PRACTICE

Early work by Mackinder (1980) focused on how building professionals make decisions regarding the

selection of materials and products. The research showed that designers were conservative in their specification practices, developing a 'vocabulary of favourite products' that once adopted were implemented on successive projects. It was observed that only items of proven reliability were continually used and designers discarded any that were associated with failure. It was found that unfamiliar items were only considered when the designer's vocabulary of favourite products did not offer a solution to a problem. Even then, designers were only persuaded to utilize new materials if their performance was supported by extensive research and offered obvious advantages.

While investigating the diffusion of innovations within the building industry, Emmitt (1997) observed similar behaviour taking place during the specification process to that previously described by Mackinder (1980). The strategy of adopting a 'palette of favourite products' is believed by Emmitt (1997) to act as a barrier against the introduction of new products unfamiliar to the designer. Furthermore, Emmitt (1997) considers the habitual use of a palette to have significant practical implications on the uptake of green items as they are seen as innovations and will therefore encounter the same resistance as any other new product to the construction industry. However, Emmitt (1997) discovered that when designers are exposed to product information for an unfamiliar item that appears favourable, it may be retained for later use. This finding suggests a state of passive awareness that may allow the adoption of alternative products and expansion of the palette. Emmitt (1997) also discusses the event of palette failure where a designer does not hold information on a product required to satisfy a specific need. In this instance, Emmitt (1997) established that designers would research and consider the use of unfamiliar products. This process could be described as a state of active selection. It may be deduced that both passive awareness and active selection could lead to the use of an item not initially contained within a designer's palette. However, the situation is subject to greater complexity as the work of Emmitt (1997) revealed that multiple palettes, such as those developed on an organizational level, are available to the designer when making a specification decision.

The work of Chick and Micklethwaite (2004) investigated the specification practices of design professionals with regard to recycled products and materials. Their study documented the obstacles to the specification of such items with lack of information identified as the largest obstacle, although not by a substantial margin. The research highlighted that the information of greatest value to specifiers would be the types of products and materials available, technical data and the location of suppliers. However, it has been reported (Peat, 2007) that technical information sources are readily available, disproving the premise that paucity of information providing instructions for the successful integration of green items is a barrier. Therefore, the residual problem is a general uncertainty about what green materials and products are available and where they can be obtained.

Emmitt (1997) recognized that professional organizations operate gatekeeping procedures that control the flow of information between product manufacturers and specifiers. This behaviour would screen out certain promotional methods such as mail shots and cold calling by trade representatives that are perceived (Architects' Journal, 1981) as annoying and a waste of time. Emmitt and Yeomans (2008) acknowledge that gatekeeping prohibits information about building product innovations from reaching the specifier. Therefore, any successful marketing campaign must be able to circumvent the gatekeeping mechanisms that restrict the specifiers' exposure to information.

Product directories and manufacturers' literature, including their websites, have been identified by the Barbour Index (1994, 2001) as the most frequently used sources of information when selecting products and materials. In addition, it has been discovered that designers often make informal contact with manufacturers that they are familiar with when faced with specification problems (Emmitt, 2006). However, product directories are only consulted during active selection, and manufacturers' literature can only be accessed once the specifier becomes aware of the product or company. Despite the growth of the Internet in recent years, trade journals are documented (Architects' Journal, 1981; Emmitt, 1997; Barbour Index, 2001) as being the main source of information with regard to new products and materials when the specifier is not actively searching. The advertisements and articles they contain may be read during a state of passive awareness to be recalled at a later date during active selection in order to fulfil a purpose, a situation described by Emmitt (1997).

This paper questions whether awareness of green materials and product suppliers needs to be raised. It has been acknowledged that designers obtain information on products and materials mainly from construction industry periodicals (Architects' Journal, 1981; Emmitt, 1997). It would appear that these publications manage to penetrate the gatekeeping procedures to become a familiar source of information to construction professionals. To test this assumption, four of the most relied upon sources of information were scrutinized to establish the extent of the promotion of construction materials with sustainable credentials.

METHODOLOGY

Four publications were selected in order to study their advertisement content and to establish the extent to which manufacturers, processors and distributors promote materials and products with sustainable credentials. Editions covering the periods of May 2005 and June 2008 were reviewed to enable a comparison to be made that would allow the identification of any change between the intervening times. Publications were chosen to represent target audiences in the architectural discipline that are available through a range of circulation methods. These included two magazines aimed at the architect and the architectural technologist, supplied by their respective professional institutions. These were deemed to be appropriate as they are received by institute members as a matter of course, thus bypassing gatekeeping procedures and without the need to be actively purchased. The most popular off-the-shelf periodical aimed at designers from the architectural profession was also included. Additionally, a control publication of specialist nature, marketed towards environmentally aware professionals, provided a check that was anticipated to contain a high proportion of advertisements for green materials and products. Details of the individual publications and their criteria for selection are described below.

Architectural Technology

The magazine is a bi-monthly publication which the circa 9000 members of the Chartered Institute of Architectural Technologists (formerly the British Institute of Architectural Technologists) receive as part of their subscription. This periodical has been chosen as it represents the information passively received by the chartered architectural technologist.

RIBA Journal

The circa 30,000 members of the Royal Institute of British Architects receive the *RIBA Journal* on a monthly basis as part of their subscription. This publication has been selected as it represents the information passively received by the chartered architect.

Building Design

Principally aimed at architectural designers, this weekly publication has to be actively purchased by choice. *Building Design* has been chosen as it is rated in the Willings Press Guide (Romeike Research, 2005; Cision UK, 2008) as the most popular periodical of its kind with circulation figures of 27,113 in 2005 and 23,836 in 2008.

Green Building Magazine (formerly Building for a Future)

A quarterly publication aimed at 'green building' professionals, *Green Building Magazine* has to be actively purchased by choice with a readership of 3500 (Cision UK, 2008). This periodical has been selected as it represents a control illustrating the type and range of advertisements targeting the environmentally conscious professional through a more specialist publication.

Issues covering the research periods for each chosen publication were studied using a data collection sheet to record advertisement details for later analysis using spreadsheets in Microsoft Excel. Only advertisements for materials, products and services were scrutinized, recruitment and event promotions were considered irrelevant to the research and were excluded. Supplements and loose flyers were also omitted on the grounds that gatekeeping procedures could screen out such information.

DATA COLLECTION

Each advertisement in the selected periodicals was examined on an individual basis to extract citation information, contact details and content data.

Citation information

This category of information consists of reference details pertaining to the periodical and the advertisement's location and format. Particulars of each publication were recorded including name, date, publisher, frequency, format and target audience. The page number and physical attributes of each advert were documented.

Contact details

The presence of the advertisers' contact information was recorded in order to demonstrate the thoroughness and credibility of the study and offer robust data.

Content data

Data in this category were used to evaluate the extent to which green building products and materials are promoted. A standard set of criteria were applied to all advertisements noting the subject advertised, whether it was shown and how it was depicted. Each advertisement was assessed to determine whether it related to a building product or material. Where this was the case, the promotion of an item's green properties was questioned, requiring a subjective decision to be made.

RESULTS

Some challenges were encountered as the subject of the advertisements was not always clear and required further investigation before complete and meaningful data could be collected. A number of advertisements, particularly those of a minimalist design, did not show or describe the item or service being promoted. Viewing the company's website or undertaking an Internet search always resolved the problem in these cases.

Tables 1, 2, 3 and 4 present the results of the advertisement examination further to the analysis of the quantitative data recorded.

FINDINGS

In order to ascertain the level of promotion afforded to green building products and materials, a total of 394 and 405 relevant advertisements were scrutinized from the 2005 and 2008 data sets, respectively.

Table 1 shows that 55% (186) of advertisements contained within the 2005 mainstream periodicals related to building products and materials. A nominal increase to 58% (201) can be seen in Table 3 for the 2008 editions. It was only possible to identify 4% (seven) and 10% (21) in the respective data sets that had green properties. Albeit small, this does indicate an increase in the number of advertisements claiming green credentials.

Figures representing the content of the control publication appear in Tables 2 and 4. Out of 56 relevant

TABLE 1 Mainstream publications, May 2005

	ARCHITECTURAL TECHNOLOGY		BUILDING DESIGN		RIBA JOURNAL		TOTAL	
	NO.	%	NO.	%	NO.	%	NO.	%
No. of adverts	2	<1	198	59	138	41	338	100
Is item a building product or material?								
Yes	0	0	83	42	103	75	186	55
No	2	100	115	58	35	25	152	45
If yes, are green properties promoted?								
Yes	0	0	3	4	4	4	7	4
No	0	0	80	96	99	96	179	96

TABLE 2 Control publication, 2005

	BUILDING FOR A FUTURE	
	No.	%
No. of adverts	56	100
Is item a building product or material?		
Yes	44	79
No	12	21
If yes, are green properties promoted?		
Yes	17	39
No	27	61

TABLE 3 Mainstream publications, June 2008

	ARCHITECTURAL TECHNOLOGY		BUILDING DESIGN		RIBA JOURNAL		TOTAL	
	NO.	%	NO.	%	NO.	%	NO.	%
No. of adverts	35	10	178	52	132	38	345	100
Is item a building product or material?								
Yes	23	66	80	45	98	74	201	58
No	12	34	98	55	34	26	144	42
If yes, are green properties promoted?								
Yes	1	4	10	13	10	10	21	10
No	22	96	70	88	87	89	179	89

TABLE 4 Control publication, 2008

	GREEN BUILDING MAGAZINE	
	No.	%
No. of adverts	60	100
Is item a building product or material?		
Yes	50	83
No	10	17
If yes, are green properties promoted?		
Yes	20	40
No	30	60

advertisements in 2005, 79% (44) related to building products or materials. From this figure, green properties were promoted in 39% (17) of cases. A marginal difference is shown in Table 4 where 83% (50) of the 60 advertisements examined were for building products or materials. The promotion of green properties occurred in 40% (20) of the 2008 marketing.

The results demonstrate a significant lack of promotion for green building products and materials in mainstream periodicals. In contrast, the specialist publication used as a control included a higher incidence of advertisements marketing green items (which was not unexpected).

Despite the environmental focus of the control publication, it was not always clear whether building products and materials advertised had sustainable credentials as the origins of the item were not communicated in all cases. This absence of clarity applied to 61% of advertisements for building products and materials in 2005 and 60% in 2008. The evidence indicates that sustainable credentials are endorsed even less in mainstream publications. It was not possible to determine whether 96% of building products or materials publicized in 2005 and 89% in 2008 had green attributes.

During the study it was noted that the sustainable credentials of items known to the researcher to be green were not always declared. This situation was common to both the control publication and the mainstream periodicals, with the greatest occurrence in the latter group. Additionally, there are a number of other issues that were observed. Some advertisements made reference to sustainability or promoted how the product could be recycled without any clear indication as to the source of the item. The limited supply of information in promotional material appears to be a technique used to entice the specifier to investigate the item in greater depth. This may be undertaken by viewing the company's website, carrying out Internet searches, examining product directories or, most desirably from the sellers' point of view, contacting the company directly. It is uncertain as to how a specifier would react to such a situation, although enquiries are more likely to be made during a state of active selection, as proposed by Emmitt (1997).

CONCLUSIONS AND RECOMMENDATIONS

Having examined the advertisement content of the most prominent construction industry periodicals aimed at designers from the architectural professions, it has been found that there is significantly less promotion of products and materials with sustainable credentials compared with 'traditional' products and materials. This comparative lack of promotion may be responsible for the widespread uncertainty about the availability of green items. The opportunity of introducing specifiers to new products via this trusted medium is clearly not exploited to the full. Manufacturers, processors and distributors must raise the profile of sustainable goods by commissioning advertisements that clearly publicize their green attributes.

Despite an increase in attention to sustainability in the general media, there was only a marginal increase in the number of advertisements for green items over the three-year period. Based on the findings of the research, there appear to be two main reasons as to why this may be the case. First, the poor quality of marketing material is an issue, something highlighted by Mackinder in 1980 and apparently still a concern. Companies are attempting to use the green agenda to their advantage by increasingly making reference to sustainability, but their claims are often non-specific and fail to communicate the sustainable origins of their goods. Second, companies are limiting their target audience by mainly promoting items with sustainable credentials in specialist publications rather than the mainstream construction periodicals, and thus not reaching the majority of specifiers.

RECOMMENDATIONS

The following recommendations are put forward to address the deficiencies in the publicity of green items and potentially induce the uptake of sustainable construction materials.

Manufacturers, processors and distributors must expand awareness of green products and materials by increasing the number of advertisements in mainstream construction industry periodicals. This would widen the target audience to include a far greater number of specifiers and design managers from general commercial practice rather than just those subscribing to specialist publications.

The content of advertisements promoting green products and materials must be comprehensive in describing sustainable credentials and advertisements must convey these attributes clearly. An improvement in the design quality of advertisements should be given some consideration in order to improve communication and increase the likely attraction to specifiers.

The implementation of these recommendations may go some way towards providing specifiers and design managers with the information they need to advance the sustainable agenda.

AUTHOR CONTACT DETAILS

Matthew Peat: Leeds Metropolitan University, School of the Built Environment, Northern Terrace, Queen Square Court, Leeds, LS2 8AJ. Tel: +44 (0)113 8127645, fax: +44 (0)113 812 1958, e-mail: m.peat@leedsmet.ac.uk

REFERENCES

Architects' Journal, 1981, 'Product information comes from magazines', in *The Architects' Journal*, 174(34), 380.

Barbour Index, 1994, *The Barbour Report 1994: Contractors' Influence on Product Decisions*, Windsor, Barbour Index.

Barbour Index, 2001, *The Barbour Report 2001: Construction Product Information – Delivery Preferences and Trends*, Windsor, Barbour Index.

Chick, A. and Micklethwaite, P., 2004, 'Specifying recycled: Understanding UK architects' and designers' practices and experience', in *Design Studies*, 25(3), 251–273.

Cision UK, 2008, *Willings Press Guide*, Volume 1, Bucks, Cision UK.

Emmitt, S., 1997, *The Diffusion of Innovations in the Building Industry*, PhD thesis, University of Manchester, UK.

Emmitt, S., 2006, 'Selection and specification of building products: Implications for design managers', in *Architectural Engineering and Design Management*, 2(3), 176–186.

Emmitt, S. and Yeomans, T., 2008, *Specifying Buildings: A Design Management Perspective*, 2nd edn, Oxford, Butterworth-Heinemann.

Mackinder, M., 1980, *The Selection and Specification of Building Materials and Components*, Research Paper 17, University of York Institute of Advanced Architectural Studies.

Peat, M., 2007, 'Barriers to the uptake of sustainable construction materials: Research outcomes', in *Architectural Technology*, 69, 25.

Performance and Innovation Unit, Cabinet Office, 2001, *Resource Productivity: Making More with Less*, London, HMSO.

Romeike Research, 2005, *Willings Press Guide*, Volume 1, Bucks, Romeike Research.

ARTICLE

Mobilizing the Courage to Implement Sustainable Design Solutions: Danish Experiences

Susanne Balslev Nielsen, Birgitte Hoffmann, Maj-Britt Quitzau and Morten Elle

Abstract

Within the built environment, stakeholders tend to implement well-known design solutions, even though sustainable alternatives exist. The key question posed in this paper is: what characterizes successful processes of implementing sustainable design solutions? In an attempt to answer the question, the work focuses on examples of successful implementation in an attempt to understand the competences required. Danish frontrunner projects are described and analysed: one case concerns the implementation of low-energy houses and another describes innovative planning processes in the water sector in order to ensure consideration of sustainable design criteria in the early phases of building projects. In the first case, the public authority succeeds in supporting design managers and other stakeholders to implement sustainable design solutions; in the second case, establishment of new multidisciplinary networks and creative work forms constitutes the outset for change. The work is inspired by the actor-network theory, emphasizing the momentum of prevailing practices, and the need to (re)develop networks to support implementation of sustainable design solutions. Conclusions point to the importance of design managers and others to develop socio-technical networks and storylines to integrate sustainability in the design and building processes. Implementation of sustainable design solutions takes more than courage; it requires key competences in catalysing network changes.

■ *Keywords* – Sustainable design solutions; competences; socio-technical networks; implementation

INTRODUCTION

Sustainable buildings represent an important issue in environmental and climate policy. There is still a huge lag between the present technological ability and the actual environmental performance of the building stock (Rohracher, 2001), which is also the situation in Denmark. A recent report from the Danish Board of Technology confirms the disturbing lack of implementation of new sustainable design solutions in the mainstream building market (such as low-energy windows, energy systems to utilize solar energy, collection and use of storm water, etc.), even though implementable sustainable design solutions exist (Danish Board of Technology, 2008).

When well-established technologies, methodologies and materials are not used in the design and building phases, it is due to reluctance from key stakeholders (Tommerup and Svendsen, 2006; Danish Board of Technology, 2008) and lack of motivation from the broad group of stakeholders, which includes dwellers, house owners, construction companies, architects, engineers and the planning and building authorities. Economic benefit is most often regarded as the key motivator, but a study about investments in energy efficient solutions show that despite their apparent profitability, investments are not made. This applies not only to the building sector, but also

earthscan
from Routledge

to firms and organizations in general (DeCanio, 1993).

Implementation of sustainable design solutions is associated with risk taking, since new practices need to be induced. This leads to calls for mobilizing the courage to implement sustainable design solutions. However, in this paper, we argue that the promotion of sustainable design solutions is more about developing new and innovative networks and strengthening certain collaboration and management competences. Current research literature is extensive when it comes to innovation strategies (e.g. Verloop, 2004) and concepts of sustainable buildings (e.g. Rooney, 2008; Vallero, 2008); however, so far, research in innovative networks for implementation of sustainable building design is scarce.

This paper contributes to the international debate on architectural, engineering and design management in relation to three perspectives. First, it contributes to the general discussions about how to speed up the transformation of huge socio-technical systems; large socio-technical systems such as the building sector are characterized by having a large momentum (Hughes, 1993). Our experiences from Denmark are that the process of transformation is still slow, although the issue of sustainable building design has been on the agenda for nearly 40 years. Different waves have characterized the development of sustainable building design, and currently we find ourselves in a phase where physical signs of climatic change have positive impacts on the willingness to initiate transformations. Second, the paper addresses how sustainability can become an integrated element of design solutions instead of being an (expensive) add-on, which is often the case today. By integrating sustainability into the process, as basic quality parameters relevant to the environment and energy, design, indoor climate and robustness to future needs, we can avoid the situation where sustainability perspective is reduced during the design process due to economic pressure. Third, we aim to explore roles and competences that support the integration of sustainability in design solutions. In order to 'turn the super tankers around' we need to consider the socio-technical complexity of the systems and the challenges, and this requires more than simply courage because competences are needed in order to induce

the necessary transformations. This perspective is of relevance for knowledge development in universities, businesses and public organizations, emphasizing the need to supplement technical skills with more generic competences, such as being able to understand how different contexts shape the development of technology, as well as to collaborate in multidisciplinary groups (Emmitt and Gorse, 2007; Hoffmann *et al*, 2009).

METHODOLOGY

The research question posed in this paper is: what characterizes successful processes of implementing sustainable design solutions? This question is addressed by exploring characteristics of stakeholder processes, where sustainable design solutions have been successfully implemented, in order to outline key challenges and the competences needed to overcome these. Specifically, the paper studies Danish projects that have empowered design managers and other stakeholders to implement sustainable solutions in the design and building phase. The answer to this research question is based on case studies of new and successful projects about innovation in building design. These projects are innovative in a Danish context and are part of larger ongoing research projects.

Eight key actors were interviewed about the process leading to the first low-energy housing area in Denmark. We used the actor-network approach, interviewing one actor and identifying other key actors by asking about their most important collaborator. We interviewed the house owners of one of the first houses, the construction company, a technology provider, the technical adviser, the former mayor, the technical manager in the municipality and two municipal planners. Insurance companies and bank associates have yet to be interviewed.

As the initial phases of a building project are one of the important phases for implementing sustainable design in mainstream building projects, we explore how the public authority supports innovative thinking among different stakeholders. This includes technological, organizational, social and institutional changes. Specifically, we studied the processes and outcome of five multidisciplinary workshops from two other ongoing research projects. There were about 17 participants in each workshop. In the research project 'Black, Blue, Green – Towards Integrated Water Management' (www.2BG.dk), the approach is to bring disciplines of urban planning and technical infrastructural planning together in a series of workshops. Hence, this project works across disciplines within municipalities. In the project '19 K – innovation in the municipal technical infrastructure' (www.19K.dk), in which more than 20% of the Danish municipalities participate, platforms are established that include sewage planners from a series of organizations including municipalities, consultancies and universities as well as professional organizations. The explicit objectives are to challenge each other and develop innovative approaches and technologies.

To understand the development of successful implementation processes we need to understand the full planning, design and implementation process and the many sub-decisions that together lead to the implementation of a sustainable design solution. For this purpose we use the actor-network theory (ANT) (Callon, 1986) to open the 'black boxes' of the implementation. This theory is most useful because it not only describes the construction of a socio-technical artefact, but also, from a design management perspective, captures maybe the most important factors – the relations between actors.

NETWORKS FOR SUSTAINABLE DESIGN SOLUTIONS

The implementation of sustainable design solutions implies not only technical innovations, but also social and organizational innovations. A conventional view of technical change, among researchers and policy makers, is that it is a one-way process of technology transfer (Shove, 1998). In this paper we argue that this technology-deterministic view of change is too one-sided and simple because it does not acknowledge the contextual complexity of system innovations. Lack of distribution of sustainable design solutions is not the result of social obstacles or non-technical barriers, but is due to the complex character of socio-technical development (Shove, 1998). It is important, but insufficient, that policy makers formulate political objectives since these alone will not entail implementation of sustainable design solutions. Such

objectives have surely contributed to raising sustainable efficiency in many regards, but some of the more critical issues are still not addressed (Sachs et al, 2000). This suggests that political objectives have certain limitations when it comes to the implementation of more challenging sustainable design solutions. Here we wish to apply a more complex network perspective in our understanding of the promotion of sustainable design solutions.

In large socio-technical systems, such as the building sector or the water sector, innovation obviously implies complex processes of change. In Denmark, there are some examples of incremental or radical sustainable changes in such systems. As an example, renewable energy has had a major breakthrough in the supply network as wind turbines changed from being ridiculed to become a source of supply and also a Danish export success (Jensen, 2003). Analysing such processes of transformation, avoiding a technology-deterministic perspective, provides rich insight into the complex processes of developing contexts, storylines, technologies, perceptions, skills, etc. (see e.g. Hughes, 1993; Geels, 2005).

A way to conceptualize processes of change in socio-technical systems is to conceptualize these as processes of reordering prevailing networks. According to ANT, socio-technical systems should be conceptualized as heterogeneous networks of both human and non-human constituents (Callon, 1986; Latour, 1987; Law, 1992). This conceptualization points to the significance of specific details in the process of change. Rather than understanding change as an overall and linear process, where one or two genius individuals have initiated changes in a system, this theory brings out the 'infrastructure' of innovative achievements. This draws attention to ever-changing network constellations in socio-technical systems, where existing relations are continually broken up and replaced with new ones. From this theoretical perspective, we interpret the process of implementing sustainable design solutions as a process of translation; meaning that certain innovators take a leading role in order to create a platform to change and stabilize new networks (Law, 1992). A classic example of this is Callon's study of the role that the Electricité de France

(EDF) played in pushing for development of the technological innovation of the VEL (véhicule electrique) in France (Callon, 1986).

Actor-network theory was used to analyse Danish projects that integrated sustainable design solutions in the building and water sectors. In the first case, we followed a typical translation process, where the municipality of Egedal has acted out as an innovator and succeeded in reordering the existing actor-network in order to support implementation of low-energy housing in an urban extension project. In the second case, we analysed the processes aimed at strengthening the role as innovator or translator by following initiatives to establish innovative platforms to implement innovative thinking. The cases are described on the basis of qualitative studies of these processes, involving a number of interviews and workshops with different actors (as described above). Some of the challenges that these actors have faced are outlined and the roles and competences necessary to achieve network changes that support sustainable design solutions are analysed. Some core elements of the process of reordering existing networks are identified, i.e. a common process of formulating problems, identify viable solutions, enrol and commit other 'actants' in the process and mobilize actual changes. Our point is to show that sustainable design solutions imply changes in existing networks, and this necessitates specific actions based on certain competences.

CASE 1: CATALYSING LOW-ENERGY DESIGN SOLUTIONS

This case analyses the urban expansion project in Stenløse, a town in the north-western suburbs of Copenhagen, as a frontrunner of sustainable design solutions in Denmark. The first inhabitants of Stenløse South have moved into their houses and during the next couple of years a total of 750 low-energy houses will be erected. In this project, the municipality of Egedal has challenged existing building practices by launching tighter local energy performance requirements than those formulated in the Danish Building Code. The local requirement is set at 35 kWh/m² per year, corresponding to the Danish category of 'Low-Energy Class 1', whereas the national minimum requirement is 70 kWh/m² per year.

As a result, 5% of inhabitants in the municipality will live in low-energy houses by 2009 (Harsvik, 2008). The initiative in Stenløse illustrates the complex work and skills needed to promote new low-energy design solutions in the building sector.

In existing building practices, a lock-in to specific levels of energy performance of new-build has been hindering new low-energy design solutions. This level of energy performance is to a great extent dictated by the minimum requirements set out in the Building Code. These requirements represent a non-human 'actor' (or stakeholder) that acts as a centre of gravity in the existing actor-network, since the building industry and its customers tend to sustain these requirements as a building standard. For example, the sales manager of Lind and Risør, one of the major Danish housebuilding companies, explains how the company will not market low-energy houses unless there is a demand from their customers (Bertelsen, pers. comm.). This illustrates how the building industry is driven by customer demands and prevailing regulations: a widely acknowledged deadlock of supply and demand in the market for sustainable buildings where neither customers nor producers take the lead, since both players are awaiting initiatives from each other (Rohracher, 2001).

The Stenløse project shows that municipalities may play an important role as a translator of new design solutions in the building sector. In this existing actor-network, Danish municipalities have somewhat of a secondary role as they merely control compliance to current regulation. Municipalities seldom initiate new energy performance building requirements themselves but merely enact existing regulation standards. However, a municipality can potentially become the 'missing link' between customers and the building industry in the mainstream market for new build. According to the planning director of Egedal municipality, the project in Stenløse is an attempt to deal with the prevailing deficiency in the mainstream market in this respect (Holm, 2006). The municipality initiated an attempt to stimulate the mainstream market to promote low-energy buildings. This initiative is based on a headstrong willingness to challenge the practices of mainstream builders and a dedicated effort to support the innovative initiatives taken in the building projects in Stenløse.

A crucial step in the process has been the development of a convincing storyline for initiating new low-energy design solutions. This storyline has given the initiative a firm direction and stability in a situation of change. The visions of one of the municipal officials, working with sustainable development of the municipality, were the beginning of this storyline. This official was frustrated about the deficiencies in the mainstream market for new building designs, and felt a need to do better than just stating political objectives of promoting more sustainable buildings in planning documents (Poulsen, pers. comm.). So far, it seems that these political objectives in local planning documents have little impact on building practices. The problem was framed as lack of adoption of existing and well-documented technologies and concepts that could sustain low-energy buildings in the building sector. The solution was framed as bringing the municipal role and authority into play with the specific intention of inducing actual changes in prevailing practices.

Viability is crucial for ensuring support for changes, since commitment from the participating actors is necessary. This is also the case in Stenløse where new design solutions have to be developed and implemented by the building industry and accepted by the customers. Being aware of this situation, the officials in the municipality developed a strategy to commit the building industry and its customers to the project. Their strategy involved preparation of a total economic analysis showing that low-energy houses are profitable in the long run. This analysis was made by a technical sparring partner, Cenergia, since the municipality did not have the necessary technical knowledge in-house. Analysis showed that by adopting well-known technologies it was possible to build a Danish standard house with low-energy status and a short payback period. This analysis became important in convincing politicians in the municipality, the building industry and its customers of the idea of implementing low-energy houses in Stenløse.

A critical point was how to get commitment from the mainstream building industry and its customers, since the activities of these actors are mainly derived from the traditional top-down regulation. The experience of the municipality was that the industry would take the easy way out in such projects and, therefore, the municipality had to think strategically about how to force the industry to adopt stricter energy performance requirements than those set in the building code (Holm, 2006). The municipal strategy was to buy land in the area, making the municipality able to enforce legally binding easements on each resold housing plot. This strategy involved somewhat of a detour compared with existing planning tools, which proved to be inefficient, since no legal requirements could be set. It necessitated a lot of effort among the officials in making themselves acquainted with planning tools. This strategy was forceful, since it compelled the buyers of building plots to comply with the energy performance requirements set out in the easements of the plot. Initially the housebuilding companies were sceptical about the easements and being 'forced' to build low-energy houses. For example, Lind and Risør initially thought that the municipality had 'scored an own goal' because the company's experience was that no customers would want to build in that area (Bertelsen, pers. comm.). Nevertheless, the municipality realized its strategy by initiating a procurement phase based on tender documents describing thoroughly the requirements and benefits regarding low-energy housing.

Commitment was not given beforehand, but required determinism and ability to convince the building industry that it was plausible to implement low-energy housing. Several potential customers showed interest in buying building lots in Stenløse during the procurement phase. These customers typically contacted different building companies in order to enquire about the company's ability to comply with the regulations in Stenløse. For example, the Andersen family wished to build a standard house produced by Lind and Risør and involved the company in their building project (Andersen, pers. comm.). This led to an extensive interchange of knowledge and ideas between the building company, its customer, its suppliers, the municipality and its technical sparring partner and national experts. During this communication, the municipality acted as motivator and catalyst, trying to suggest solutions and to overcome barriers in the process. The outcome was a model of a marketable standard low-energy house by Lind and Risør. Similar processes took place with other building companies and their customers. The municipality has now integrated the requirements from Stenløse South into the municipality as a whole.

The Stenløse case shows how the municipality of Egedal has drawn upon different competences in the process of mobilizing changes in prevailing building practices in order to develop new design solutions (Eliasen, pers. comm. and Madsen, pers. comm). The buildings in Stenløse would not have been low-energy buildings if the municipality had not pushed for these new design solutions. Experience from the project shows that it requires a lot of effort to change current building designs within mainstream building companies. Through the involvement of the municipality, it was possible to support new low-energy design solutions. In order for this support to be successful, the officials in the municipality had to develop new skills and competences in planning and networking for innovations. They had to learn about committing other actors, developing viable concepts and build up networks. It required determinism and complex work to carry this process through. The successful outcome of the process in Stenløse owes much to politicians and officials in the municipality of Egedal and their visions, strategies and abilities. A key strategy was to support concrete collaborations between the municipality and the building industry and their customers. In the next case we discuss how this process of building innovative networks may take other forms.

CASE 2: CATALYSING INNOVATION IN MULTIDISCIPLINARY TEAMS

Climate change has to be addressed not only at building level, but also at a larger urban scale. In the summer of 2007, Denmark experienced extensive flooding of houses and streets, resulting in a call for radical changes in the design of the built environment, in order to prevent flooding and to maintain the value of houses. Providing sustainable urban infrastructure is a part of implementing sustainable design; however, this is not easy, because of the momentum and a resistance to

change in the way urban infrastructure is managed (Nielsen and Elle, 2000).

Multidisciplinary networks are needed for innovation in large socio-technical systems. One challenge is to establish platforms for meetings between central actors who can support the creation and implementation of innovations through the creation of new storylines, networks, competences, etc. This case also emphasizes non-human 'actors' (or stakeholders) such as climate change with heavy rain, storms, surges, floods and oxygen depletion as this has focused public attention on the sewage sector in Denmark. Increasing intensity of rain and lack of capacity in the established infrastructure, causing incidents of overflow and flooded buildings, have resulted in rapidly growing motivation to find new ways of reducing the amount of storm water entering the drainage system and the treatment plants. The need for investment is another issue negotiated among researchers, politicians, real estate agents and house owners. While the government has not yet put forward its analysis of the potential problems, researchers in alliance with the press warn that the loss of value for houses in threatened areas in Denmark has to be taken into account and heavy precautions have to be taken. Even though the prices of houses in the flooded areas have stabilized, a storyline of risk and investments is certainly developing.

The conventional response to lack of capacity is to add more capacity, meaning larger pipes and more retention basins on municipal ground. Landscape-based handling of storm water in urban areas is an example of a new design approach, which needs a different network than the conventional sewage system. The landscape-based handling of storm water further aims at integrating retention capacity in the urban landscape in a way that adds biodiversity to an urban area and enhances the human experience of an attractive urban area. But handling storm water, using the city parks, streets, squares, stadiums, etc., needs collaboration between different departments within the municipal organization, especially the urban planners, the sewage planners and the road and park administrations. Hence, the challenge for this approach is to gather relevant actors around a new storyline – 'storm water handling as part of creating attractive urban environments' – and to ensure that consideration of infrastructure is included in the early stages of an urban planning process. This is the objective of one of the two projects presented here. The project aims at innovating beyond the pattern of conventional planning and development by establishing multidisciplinary teams with different competences. These can be labelled as a sort of virtual R&D department, and here we point to different ways of advancing such departments in the large socio-technical system of sewage planning.

As the sewage sector is a large and very complex socio-technical system, not surprisingly, most innovation seems to take place within parts of the system. Important innovations can be identified, such as the development of municipal wastewater treatment plants in the 1970s and 1980s. However, in order to enable radical innovation, holistic perspectives and coordination are needed. This includes the actors' ability to understand each others' perspectives and tasks, and to engage in developing possibilities for action.

The basic element of both projects is to introduce new translation processes in the sector in which different actors meet to exchange perspectives and to work together. Not merely to develop new designs, but also to introduce and perhaps take these directly into use in their professional practices. A central action is that the different actors work together to agree on common understandings of the challenges of the sector to be met and worked on. Traditionally, the sector frames a series of different networks existing side by side, producing and reproducing different problems and solutions and hence, directions (perhaps counteracting or delaying each other). The collaborations provide common understandings with an immediate momentum as more people agree. However, the step is of more radical importance. A series of new networks are constructed around the challenges formulated and, in future phases, the actors are expected to develop these networks around new solutions including concrete collaborations and pilot projects. Furthermore, the projects frame a more systematic methodological approach to promote creativity and reflection among the participants.

Negotiations on investments are central. An interrelated challenge formulated by the actors is the current liberalization taking place in the water sector. The municipal suppliers are being sold off as independent municipal companies: on the one hand, this separates the planning departments and the water sector in the municipality, but on the other hand, it makes the need to establish coordination very visible which was seldom seen before. The new water companies are benchmarked on their productivity and, together with the need for investments, this calls for new business models as important parts of the innovative strategies.

This approach to innovation in the large socio-technical systems underlines certain skills that eco-preneurs need (Schaper, 2002). The set-up of the projects has been ingeniously launched based on motivating storylines combining current societal spotlights, i.e. the urge to develop user-driven innovation as part of being an innovative society and the need to meet the consequences of climate change and the trend of developing new urban qualities.

BREAKING OUT OF THE WELL-KNOWN

Even though we have studied successful cases of implementation of new sustainable design solutions, our cases confirm the findings of the Danish Board of Technology (2008) when it comes to barriers for sustainable building design. There is a general resistance that is reflected by reactions like: what is it good for? Is it necessary? Is it worth it? Is it technically and economically viable? Is it a good idea in the short and long term? And who will benefit and who will bear the costs and the risks?

In addition to the barriers identified by the Danish Board of Technology (2008), our research has identified a lack of planning and management processes which support the construction of new networks. Implementing new design solutions may take some kind of courage, but the above cases of innovation in the built environment show that other competences are needed in order to implement sustainable design solutions.

A key characteristic of a successful process is acknowledgement of the complexity of a design solution framing socio-technical aspects and the need to redefine one's roles and obligations. In the case of low-energy housing, the municipality provided the obligatory local plan, but it also engaged in the market as well as in dialogue with standard house companies, future dwellers and providers of building materials. In this way, it established a new actor-network which together could realize this ambitious project. In the case of storm water handling, it is equally recognized that several planning professions need to collaborate at an early stage of an urban project in order to ensure that area-based storm water handling is an attractive design solution. Implementations at later stages will include participation from many others and especially building owners, the users of the urban space, politicians and health and safety authorities. Along this line is also the discussion of who should have the innovation roles, the state, the municipality, private investors or the building companies?

Visionary framing of problems and future scenarios is another characteristic of a successful process. In the above cases, there have been persons in the decision-making process who have seen possibilities of doing something differently and have been able to include others in their visions. In the case of low-energy housing, the first translator was an official, later it was a small group at the town hall and now, when the network is expanded and the project is seen as a success, other actors in the network like the producers of building materials can be seen as translators of low-energy design solutions into other building projects and networks. In the case of integrated storm water management, it is different professions within the water sector that are engaging in the innovation of existing practice, through common reflections on their own role, confrontation of their 'own world' and the socio-technical infrastructure system as a whole. The approach in both research projects was to create platforms for reframing visions and exploring possibilities for new strategies and practices, and the participants can act as translators in their own organizations.

To choose new design solutions should not merely be perceived as risk taking, and having the courage to do so. It also takes the competences to formulate a convincing storyline, to connect and commit other actors, and to mobilize technological, organizational, social and institutional changes. The gains of being innovative should in our opinion also be embraced. In our research, actors expressed how breaking out of the safe and well-known provided a thrill, inspiration

and engagement in their job (Eliasen, Jensen, C.B., Jensen M.B., Madsen, pers. comms.).

CONCLUSION

We have discussed the implementation of sustainable design solutions as a complex process of reordering existing networks. Through our cases we are able to posit that network changes cannot rely on courage alone, but that innovators are also required to act as catalysts to reorder the processes. It takes motivation to gain insight into the state-of-the-art of relevant design alternatives including practical and theoretical knowledge about possibilities and implications. But just as important is the capability to see other actors' needs, to communicate with a series of different stakeholders and to facilitate processes for establishing new shared understandings and building new networks for the implementation of sustainable and innovative design solutions.

To speed up the processes of innovation in the building sector, the cases confirm that on the one hand a single visionary person can have an important influence as an initiator of a new design, but on the other hand, innovative design processes can be supported by organizing a strategic set of competences. Furthermore, in the process of promoting sustainable design solutions, public authorities can play a crucial role as translators between existing and new networks, due to their role and influence on the early phases of building projects.

Further research is needed to learn more about innovation strategies in sustainable building design and to identify ways of building new innovative networks. Our case studies point to the need of building new supporting networks and further research could help the understanding of reorganization processes in theory and practice. The perspectives of this research could change our thinking about sustainable innovation in the built environment and the sustainable design manager as a potential network facilitator.

ACKNOWLEDGEMENTS

We acknowledge informants, who have willingly shared information with us in interviews, participants from the different projects (www.2bg.dk and www.19K.dk) and colleagues at DTU Management Engineering, Technical University of Denmark.

AUTHOR CONTACT DETAILS

Susanne Balslev Nielsen (corresponding author), **Birgitte Hoffmann, Maj-Britt Quitzau** and **Morten Elle**: DTU Management Engineering, Technical University of Denmark, Produktionstorvet, Building 424, DK-2800 Kgs Lyngby, Denmark. Tel: +45 4525 1525, fax: +45 4593 3435, e-mail: sbni@man.dtu.dk

REFERENCES

Callon, M., 1986, 'The sociology of an actor-network: The case of the electric vehicle', in M. Callon *et al* (eds), *Mapping the Dynamics of Science and Technology*, London, The Macmillan, 19–34.

Danish Board of Technology, 2008, *Klimarigtigt Byggeri – Vi Kan Hvis vi Vil! (Environmentally Friendly Building in Practice – What are we Waiting For?)*, Copenhagen, Teknologirådet, www.tekno@tekno.dk

DeCanio, S.J., 1993, 'Barriers within firms to energy-efficient investments', in *Energy Policy*, 21(9), 906–914.

Emmitt, S. and Gorse, C.A., 2007, *Communication in Construction Teams*, Oxford, Spon Research, Taylor & Francis.

Geels, F., 2005, 'Co-evolution of technology and society: The transition in water supply and personal hygiene in the Netherlands (1850–1930) – a case study in multi-level perspective', in *Technology in Society*, 27(3), 363–397.

Harsvik, M., 2008, 'Egedal Kommune vil være energiby' ('The municipality of Egedal wishes to become an energy city'), in *Lokalavisen UgeNyt*, 23 September, 21.

Hoffmann, B., Jørgensen, U. and Bregnhøj, H., 2009, 'Intercultural competences and project based learning in the engineering curriculum', in M. Blasco and M. Zoelner (eds), *Teaching Culture*, Copenhagen, Nyt fra Samfundsvidenskaberne (in press).

Holm, R., 2006, 'Stenløse stiller krav til miljøkrav til nybyggeri' ('Stenløse sets sustainable demands in new build'), in *Nyhedsmagasinet Danske Kommuner*, 15, 18–20.

Hughes, T.P., 1993, 'The evolution of large technological systems', in W.E. Bijker *et al* (eds), *The Social Construction of Technological Systems – New Directions in the Dociology and History of Technology*, The MIT Press, Massachusetts, US, 51–82.

Jensen, I.C., 2003, 'Men against the wind – a Danish industrial fairy tale', Copenhagen, Børsen.

Latour, B., 1987, *Science in Action: How to Follow Scientists and Engineers Through Society*, Milton Keynes, Open University Press.

Law, J., 1992, *Notes on the Theory of the Actor Network: Ordering, Strategy and Heterogeneity*, www.lancs.ac.uk/fass/sociology/papers/law-notes-on-ant.pdf (accessed 10 October 2008).

Nielsen, S. and Elle, M., 2000, 'Assessing the potential for change in urban infrastructure systems', in *Environmental Impact Assessment Review*, 20(3), 403–412.

Rohracher, H., 2001, 'Managing the technological transition to sustainable construction of buildings: A socio-technical perspective', in *Technology Analysis and Strategic Management*, 13(1), 137–150.

Rooney, E.A., 2008, *Green Homes – Dwellings for the 21st Century*, Pennsylvania, US, Schiffer.

Sachs, W., Loske, R. and Linz, M. (eds), 2000, *Greening the North – A Post-industrial Blueprint for Ecology and Equity*, London, ZED Books.

Schaper, M., 2002, 'The essence of eco-preneurship', in *Greener Management International, Theme: Environmental Entrepreneurship No 38*, Sheffield, Greenleaf Publishing, 26–30.

Shove, E., 1998, 'Gaps, barriers and conceptual chasms: theories of technology transfer and energy in buildings', in *Energy Policy*, 26(15), 1105–1112.

Tommerup, H. and Svendsen, S., 2006, 'Energy savings in Danish residential building stock', in *Energy and Buildings*, 38, 618–626.

Vallero, D., 2008, *Sustainable Design – The Science of Sustainability and Green Engineering*, US, Wiley.

Verloop, J., 2004, *Managing Innovation by Understanding the Laws of Innovation*, Amsterdam, Elsevier.

PERSONAL COMMUNICATIONS

Andersen, B. and Andersen, K., Family building in Stenløse South, personal communication (28 June 2007).

Bertelsen, M., Lind and Risør A/S, personal communication (5 July 2007).

Eliasen, W., Municipality of Egedal, personal communication (29 March 2008).

Jensen, C.B., Vilhelm Lauritzen Architects, personal communication (18 September 2008).

Jensen, M.B., University of Copenhagen, personal communication (January – October 2008).

Madsen, J., Municipality of Egedal, personal communication (30 August 2008).

Poulsen, J., Municipality of Egedal, personal communication (26 June 2007).

ARTICLE

A Design Process Evaluation Method for Sustainable Buildings

Christopher S. Magent, Sinem Korkmaz, Leidy E. Klotz and David R. Riley

ABSTRACT

This research develops a technique to model and evaluate the design process for sustainable buildings. Three case studies were conducted to validate this method. The resulting design process evaluation method for sustainable buildings (DPEM[SB]) may assist project teams in designing their own sustainable building design processes. This method helps to identify critical decisions in the design process, to evaluate these decisions for time and sequence, to define information required for decisions from various project stakeholders, and to identify stakeholder competencies for process implementation.

■ *Keywords* – Sustainable design; design process; decision-based design; process modelling; integrated design

INTRODUCTION

As the demand for 'sustainable' or 'green' buildings continues to increase, the need for a better understanding of how these buildings are designed has also increased (USGBC, 2008a). Sustainable building projects often require integrated design approaches to perform complex design analyses, energy modelling and system optimizations (Riley *et al*, 2004). There is broad recognition that these buildings demand increased levels of design integration between structural, envelope, mechanical, electrical and architectural systems (e.g. Reed and Gordon, 2000; Kratzenbach and Smith, 2003; Lewis, 2004). Therefore, interpersonal skills become more critical in these projects compared with traditional ones (BHKR, 2003).

However, the design process for sustainable buildings is largely undefined and is reinvented on a project-by-project basis. As teams of highly specialized and fragmented disciplines are formed for a particular project, a new design process is developed. As a result, a highly sequential and specialized business practice has evolved for building design teams, which challenges efforts for collaboration and integration.

The current practice in sustainable building shows that the key roles of early stakeholder involvement, collaboration and integration for sustainable building design are recognized (Reed, 2003). Still, the specifics of the sustainable design process for individual projects remain unclear. In response, this research utilizes a proposition-based case study approach to develop and validate a method that design teams can use to help them plan their design processes for sustainable buildings. Six propositions, based on background from theory and practice, were evaluated in this study on three separate case study projects. Five of the six propositions were validated and these validated propositions comprise the design process evaluation method for sustainable buildings (DPEM[SB]). This iterative method enables identification and evaluation of critical design process decisions, definition of required information, and identification of required stakeholder competencies.

BACKGROUND

The rise of specialized disciplines in the 1970s resulted in a traditional design process that isolates disciplines during the design and construction process (Kashyap *et al*, 2003). This process is similar to the 'over-the-wall' approach in manufacturing design (Evbuomwan and Anuba, 1998). The traditional, specialized and sequential design process

ARCHITECTURAL ENGINEERING AND DESIGN MANAGEMENT ■ 2009 ■ VOLUME 5 ■ PAGES 62–74
doi:10.3763/aedm.2009.0907 © 2009 Earthscan ISSN: 1745-2007 (print), 1752-7589 (online)

is divided with milestones such as schematic design, design development and construction documents (Haviland, 1994) and can inhibit the integration among various disciplines that best serves sustainable building projects. A different design process is critical to the successful implementation of these projects. However, the design process for sustainable buildings remains mostly undefined and is reinvented on each new project.

Metrics and standards for the final building product such as the Building Research Establishment Environmental Assessment Method (BREEAM) and the Leadership in Energy and Environmental Design (LEED®) rating systems are commonplace in the green building market (BRE, 2007; USGBC, 2008b). These green building assessment systems primarily evaluate building features such as energy performance and materials used; however, they provide minimal guidance on design and construction processes to help project teams achieve these standards (Korkmaz et al, 2007). Few methods exist to evaluate the sustainable building design process, despite calls for more process-oriented sustainable building assessment (Kaatz et al, 2006). Failure to systematically evaluate the sustainable building design process can produce design process waste, which can lead to suboptimal performance of the building.

DESIGN PROCESS FOR SUSTAINABLE BUILDINGS

Among the various definitions for sustainable buildings, Raynsford's (2000) statement addresses sustainable building as 'a part of a sustainable development which aims to deliver built assets that enhance quality of life and offer customer satisfaction, offer flexibility and the potential to cater for user changes in the future, provide and support desirable natural and social environments, and maximize the efficient use of resources'. Cross-disciplinary teamwork early in the design process is essential for achieving the successful integration of building, community, natural and economic systems for sustainable development (Reed and Gordon, 2000). The success of the project is largely influenced by interactions within the team throughout the design process, thus interpersonal skills become more critical than in the traditional process (BHKR, 2003).

DESIGN DECISIONS AND WASTE

Waste, defined as activities unnecessary for task completion and value generation, is common in design processes (Koskela and Huovila, 1997). Recognizing the most prevalent forms and causes of waste is a critical step in efforts to improve the design process. Causes of waste in the design process can fall into the following three categories:

- missing design competencies: The presence of key design competencies is especially critical for sustainable projects, which require additional and greater distribution of functional competencies among team members (Reed and Eisenberg, 2003). Lack of relevant competencies during the design process will decrease the project's chances of success (Riley et al, 2004; Lapinski et al, 2006). For example, failure to include an energy expert in the early design decisions may result in missed opportunities to minimize energy use.
- poor timing of decisions: A classic tension exists in the timing of design decisions. The delaying a decision allows for gathering additional information and performing analyses, which may lead to a better decision. However, if other decisions depend on the results of earlier decisions, a cost can be associated with this delay. Developing a mechanism to evaluate the timing of decision making in the sustainable building design process may help identify and reduce waste associated with ill-timed decisions (Magent, 2005).
- missing information for decisions: Decisions made without sufficient information can lead to process waste including changes in design decisions and the breaking of commitments on which others have relied (Magent, 2005).

DESIGN DECISION NETWORK

The notion that design is a decision-making process is consistent with the definition of a decision as a choice from among a set of options and an irrevocable allocation of resources. Decision-based design (DBD) seeks to base engineering design decisions on information obtained from a variety of sources going well beyond the engineering disciplines. The preferred decision is the option whose expectation has the highest value. Classical decision theory applies to the case where a known set of options has been defined, and design is a decision that seeks to maximize value (Hazelrigg, 1998).

Decision-based design is a term coined to emphasize a different perspective from which to develop methods for design. The principal role of a designer in DBD is to make decisions, and decisions help bridge the gap between an idea and reality. In DBD, decisions serve as markers to identify the progression of a design from initiation to implementation to termination (Mistree and Allen, 1997).

The body of knowledge on making rational design decisions from a performance point of view originates from ideas presented by Herbert A. Simon in *The Sciences of the Artificial* (Simon, 1969). Many models of the engineering design process have been created. A good overview of some of these models is provided by Cross (1994) and Birmingham *et al* (1997). As most authors indicate, no universally accepted model of the design process has emerged from these studies. The heightened importance in collaboration between disciplines in the integrated design process raises the need to develop a common design process vernacular for both engineers and architects. In general, engineering models of the design process are more linear, prescriptive and tree-like, having a well-defined sequence of stages, resting on an exhaustive evaluation of requirements, and basically deal with a well-defined problem. Architectural process models tend to be more cyclical, descriptive and lattice-like, allowing for many process cycles, based partly on implicit and changing requirements and relying on tacit knowledge (De Wilde *et al*, 2002).

PROBLEM STATEMENT AND THE NEED

Understanding the design process as a network of decisions reveals the fact that current models of the building design process provide little more than a coarse set of milestones, broad categories of design decisions, and sequential building systems analysis. While the need for an integrated design has been established in sustainable buildings (Reed, 2003), the actual design process for these buildings is undefined. In practice, newly formed design teams

incur process waste that detracts from the design analysis and final building design. A better definition of the design process for sustainable buildings that stresses the integration between systems and accounts for the key processes and competencies vital to appropriate design decisions could help reduce design process waste and increase design quality.

In various forms, modelling has been applied to design processes for traditional (Austin, 2001; Tzortzopoulos *et al*, 2005) and sustainable buildings (Lapinski *et al*, 2006; Klotz *et al*, 2007). Applying a decision-oriented process and competency-based method to evaluate the design process can provide a descriptive model of the integrated design process, which is especially useful for sustainable buildings as it can identify process waste and opportunities to increase design quality. To respond to this need, this paper presents a design process evaluation method (DPEM[SB]) for sustainable buildings. A critical component of this method is the ability to model or graphically represent the design process being evaluated.

RESEARCH METHOD

To develop the method, the researchers conducted a comprehensive literature and industry practices review. Topics considered in the literature review were lean manufacturing, concurrent engineering, options-based theory and decision-based design. This review combined with information gathered from meetings with practising design experts in the field of sustainable buildings provided the basis for the theoretical design model. A round-table meeting of key industry and academic professionals from the US played an important role in this research effort. The round table was assembled in Tarrytown, New York State, in 2004 to identify current industry practices in the area of sustainable building design. The meeting developed a shared understanding of the integrated design process and produced a research agenda supporting a transformation of the building industry which also helped develop the method.

Building on the initial insight from the round-table meeting, multiple case studies were evaluated to test and refine the DPEM[SB]. Qualitative social science practices have been utilized in the form of proposition testing on multiple case studies (Yin, 2003) as a means to develop an evaluation method for the design process of sustainable buildings. The identification of supporting case study events was used to validate the research propositions. Case studies are appropriate due to the exploratory and qualitative nature of this study and because the research focuses on contemporary behaviour and does not require or allow for control of behavioural events (Yin, 2003).

FORMAT OF THE PROCESS AND COMPETENCY-BASED MODELLING TECHNIQUE

This section describes the format of the process and competency-based modelling technique developed for this research. The modelling technique is one component of the DPEM[SB] and is designed to enable the representation of prevalent sources of waste in the sustainable building design process. Key aspects of the modelling technique are described along with their representative icons.

Representing design decisions

The model needs to represent the nature of the design decision process where options are narrowed, through analysis and information, to a point where a commitment to one choice is made. Therefore, the cone icon is used to represent design decisions (Figure 1). The cone's diameter represents ambiguity in the decision process. As new design options are presented for consideration, ambiguity is large. As information is gathered and analysis of options takes place, ambiguity shrinks. Ultimately, a decision is made, which results in certainty. Since changes to decisions can create waste, such as rework, decisions are ideally made with sufficient certainty to be considered commitments upon which subsequent decisions can rely.

In addition to representing the nature of the design decision process, the cone icon can also emphasize decisions with a greater impact than others on project costs or performance. For instance, building orientation and configuration typically has a greater impact on daylight performance than sun shading design. By varying the sizes of the cone icons, the process and competency-based modelling technique allows prioritization of certain design decisions (larger icons) above others (smaller icons).

Representing timing of decisions

Within the cone icon, the model can represent timing of decisions (Figure 2). Level II decisions are those made at the optimum time. Ambiguity is sufficiently

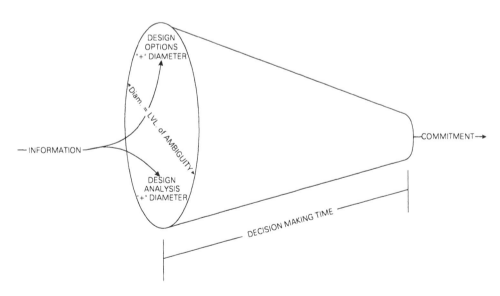

FIGURE 1 Design decision model

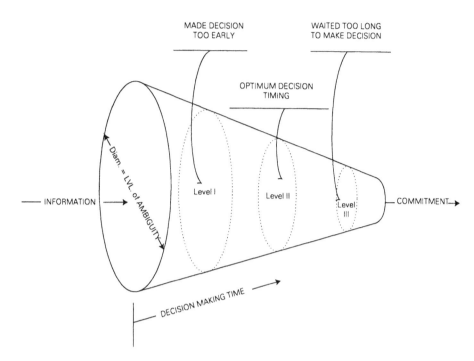

FIGURE 2 Design decision timing model

decreased to allow for an appropriately informed decision and additional waste has not been introduced to the decision-making process by prolonging the timing of the commitment. In terms of traditional decision-making theory, such as Real Options Theory, a level II decision is equivalent to the point at which the marginal benefit of making the decision is equal to the marginal cost of waiting to make the decision (Neufville, 2001). Ideally, sustainable design teams will make level II decisions, as decisions made before or after this level can result in product or process waste.

Decisions made too early with insufficient information are classified as level I decisions. These decisions can cause waste where inappropriate commitments are made, which lead to misinformed decisions. For example, sizing mechanical equipment prior to adequately understanding heating and cooling loads can lead to waste in the form of excess safety factors that are carried throughout the duration of design. Mechanical equipment that is not optimally sized would result in a building that costs more and uses more energy than needed.

Contrasting with the hasty level I decisions, level III decisions are those made after the optimum time to make a decision has passed. These late decisions are also a source of waste, primarily in the form of project delays.

Representing design competencies

Design activities occur in the design environment, which includes not only the decisions and analyses made when developing the design, but also the stakeholders responsible for the design. The set of competencies (e.g. skills and knowledge) possessed by these stakeholders directly influences the decisions, information and analyses that constitute the design process. These competencies are represented by a hexagon icon in the process and competency-based modelling technique (Figure 3).

Representing information for decisions

Organizing decisions in an appropriate sequence can reduce design process waste. A conceptual level network of design decisions is shown in Figure 3. Decisions result in commitments, which act as

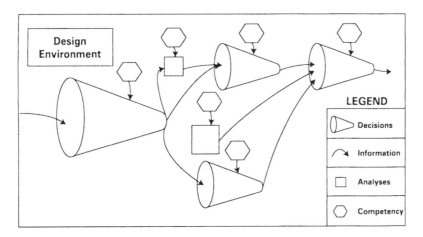

FIGURE 3 Design decision network model

information for future design process decisions. In addition to decisions, analyses such as computer simulations, cost estimates and product research are performed during the design process. These analyses are enablers of decision-making processes and are represented by boxes in the process and competency-based modelling technique.

PROPOSITIONS

Due to the exploratory nature of this research and the realities of researchers' access to sustainable design process information, a proposition-based case study approach was deemed most appropriate for developing and testing the evaluation method. The initial step in this approach involves using theory and observations to develop propositions for investigation. Then, the propositions are tested on case-study projects. Specific industry examples from sustainable building design were combined with established theory to satisfy the first step in proposition development. These propositions are listed below and the theory underpinning each is provided in Table 1. The process and competency modelling technique is an integral part of the DPEM[SB] development (Figure 4).

CASE STUDIES

The propositions comprising the design process evaluation method were investigated on three recently completed buildings. For each project, researchers studied the network of decisions,

information, analyses and competencies that comprised the design process. Due to the prominent role of energy-saving systems in the sustainable building design process, researchers focused on the energy system for these cases. Energy systems that impact the final energy performance of the building include mechanical, electrical, lighting and building envelope systems.

Intentionally, the cases selected for examination represent a range of project types, sizes and delivery systems. Despite their differences, each case emphasized energy efficiency during design, targeting a reduction in energy consumption of at least 40% compared with a standard new building of the same type. Each case also included accessible design process information and design team members.

Case study no 1 – Pennsylvania Department of Environmental Protection (DEP) Cambria Office Building

The first case study project investigated in this research was a 34,500 square foot office building housing the Commonwealth of Pennsylvania's Department of Environmental Protection (DEP) management and field personnel. Design activities began in 1997 with building construction completed in the autumn of 2000. For this building, the DEP incorporated sustainable building features (such as efficient wall and roof insulation, high performance windows, ground source heat pumps, an underfloor

TABLE 1 Propositions of the DPEM[SB] and their theoretical background

PROPOSITION	THEORETICAL BASIS
1) Building function Engineering the design process based on the building's desired functions will add value to the design process	Lean production – identifying value dictates process (Womack and Jones, 1996)
2) Design decision network The design process is a network of decisions producing commitments, connected by discipline specific analysis activities performed by design actors with associated competencies	Decision-based design – 'design decisions are based on information obtained from a variety of sources, going well beyond the engineering disciplines' (Hazelrigg, 1998)
3) Decision timing and sequencing Changes to the sequencing and timing of key decisions, dictated by the building's desired functions, can add/detract value to/from the overall design process	Design decision network model (DDNM) Integrated functional design (Evbuomwan and Anuba, 1998)
4) Pull-driven information Value is added to the design process when critical decisions that need to be made pull the information and analyses required to reach an informed commitment	Lean production – pull-driven processes (Womack and Jones, 1996)
5) Competency-based value Effective implementation of the integrated design process is dependent upon inclusion of key individual and team competencies	Team selection – 'teams should be determined by the skills needed to accomplish the work' (Luecke, 2004)
6) Decision-based evaluation model A process-based methodology to assess sequencing and timing of key decisions in the design process results in an improved design process based on value added/value lost	Lean production – continuous value enhancement through process management and improvement (Liker, 2004)

distribution system, energy recovery ventilators, daylighting and motion sensors, and an 18.2 kW photovoltaic system for on-site electricity production) into its design process from the beginning to meet its goal of producing a high-performance, environmentally friendly building.

Case study no 2 – The Pennsylvania State University School of Architecture and Landscape Architecture (SALA) Classroom and Office Building

The second case study project investigated in this research is a four-storey building housing Penn State's School of Architecture and Landscape Architecture. The 145,000 square foot building includes design studios, faculty and administrative offices, a library, computer lab and workshops. Design activities started in 2001 and building construction was completed in the summer of 2005. A sustainable design charette was conducted at the beginning of the project in part to focus the efforts of all team members on reducing the final energy consumption of the building.

Case study no 3 – The American Indian Housing Initiative (AIHI) Early Childhood Learning Center

The third case study project investigated was an Early Childhood Learning Center (ECLC). The building is a 4800 square foot facility built for Chief Dull Knife College in Lame Deer, Montana. Building construction was completed in the summer of 2007 after design activities began in 2004. The design team consisted primarily of faculty and graduate students in architecture, architectural engineering and landscape architecture.

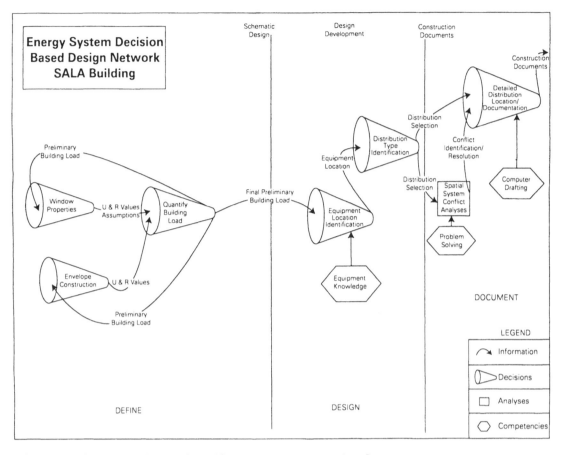

FIGURE 4 Decision-based design network model for energy systems – case study no 2

ANALYSIS OF CASES

Each of the three cases was analysed using the approach described in this section, which is based on Yin's (2003) case study approach.

Data gathering

Literature concerning the case was reviewed. Researchers identified participating companies, key project stakeholders within these companies, the stakeholders' disciplines, and the teams and structures in which the stakeholders operated. Special attention was paid to events related to the research propositions and interdisciplinary interaction during the design process. Participants in the energy systems design process were interviewed regarding the decisions, commitments, information and analyses that constituted the design process. The interviews were conducted in accordance with the general guidelines for interviews as described by Babbie (2004).

Content discovery

Audio recorded interviews in digital format were transcribed to identify project events related to the research propositions. The transcribed data were reviewed and potential events identified. Identified events were analysed to determine whether they supported the proposition. A list of confirmed events is provided in Table 2.

Event/proposition support analysis

An event could either support the given proposition in a positive or negative manner or contradict the proposition. Positive support maintained the

TABLE 2 Proposition validation results

PROPOSITION	DESIRED LEVEL OF CERTAINTY	LEVEL OF DIFFERENCE IN RIVAL THEORIES	REQUIRED EVENTS	TOTAL CORROBORATED EVENTS	CAMBRIA EVENTS		SALA EVENTS		AIHI EVENTS	
					SUPPORTING	CORROBORATED	SUPPORTING	CORROBORATED	SUPPORTING	CORROBORATED
1	Low	Little	3	7	10	3	4	2	4	2
2	Medium	Little	4	4	3	3	1	0	1	1
3	High	Little	5	9	15	4	17	2	13	3
4	Medium	Great	5	6	11	3	8	2	11	1
5	High	Great	6	8	22	3	15	3	16	2
6	Medium	Little	4	0	0	0	0	0	2	0

SALA – School of Architecture and Landscape Architecture

AIHI – American Indian Housing Initiative

proposition as stated through direct conformity, resulting in at least one identifiable value-added, stated outcome of the given proposition. Negative support maintained the proposition through non-compliance of the proposition that resulted in process waste. Both types of support are included in the results reported in the 'supporting' columns of Table 2.

Event corroboration

Interviews were performed with the design manager, architect of record (i.e. design documents coordinator) and mechanical engineer on each case study project to enhance certainty in the events identified. Corroboration of events occurred when a minimum of two sources described the same event on the same case in the same manner. As shown in the 'Supporting' and 'Corroborated' columns of Table 2, many supporting events were not corroborated. Only corroborated events contribute to validating the research propositions.

Event replications

The validity of each research proposition is determined by 'replication', which is the total number of corroborated events for all three cases. The number of corroborated events required to validate each proposition varies, depending on the criteria established for certainty in each proposition (Yin, 2003). The desired level of certainty and the level of difference in rival theories were the primary criteria established for this research. The ratings in the 'desired level of certainty' column in Table 2 correlate with the level of detail of the propositions. For instance, propositions 3 and 5 are the most detailed and, therefore, receive a 'high' level of certainty ratings. The 'level of difference in rival theories' ratings in Table 2 are classified as either 'little' or 'great' based on the presence of theories that conflict with the proposition. For instance, proposition 4 receives a 'great' rating because it is based on pull-driven theory, which has a strong opposition from push-driven management. Using the standards of Yin (2003), the ratings for these criteria establish the number represented in the 'required events' column of Table 2. To validate a proposition, the total number of corroborated events must exceed this number of required events.

RESULTS

Propositions 1–5 are validated by the case study analyses as the total corroborated events exceed the number of required events (Table 2). Proposition 6, 'A process-based method to assess sequencing and timing of key decisions in the design process results in an improved design process based on value added/lost,' was not disproved. However, the validation requirements were not met for this proposition due to a lack of evidence from the case studies selected. The validated propositions show that:

- engineering the design process based on the desired functions of the building adds value to the design process.
- the design process is a network of decisions producing commitments, connected by discipline-specific analysis activities performed by design actors with associated competencies.
- changes to the sequencing and timing of key decisions, dictated by the building's desired functions can add value to the overall process.
- value is added to the design process when critical decisions that need to be made pull information and analyses required to reach an informed commitment.
- realization of value generation from the implementation of the integrated design process is dependent upon inclusion of key individual and team competencies.

THE DESIGN PROCESS EVALUATION METHOD FOR SUSTAINABLE BUILDINGS

The DPEM[SB] represents each validated proposition as one step in a cyclical evaluation process, meant to continuously improve design of the sustainable building design process (Figure 5). This method is to communicate the research findings in a format applicable for use in industry practice.

Written descriptions supplement the graphical representation of the DPEM[SB]. The initial step in building design is assembly of the team. Core competency requirements of team members participating in the design of a sustainable building must be considered. Systematic analysis of the design team helps identify individuals with the

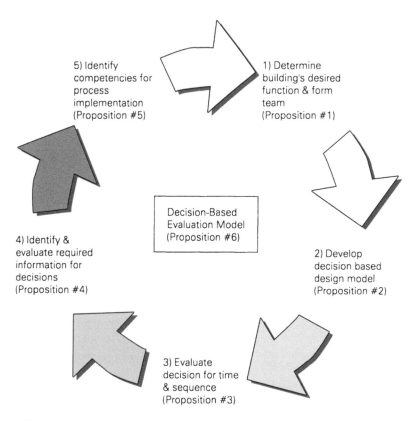

FIGURE 5 Design process evaluation model for sustainable buildings

specific competencies required to inform and make each key decision. If the team lacks a desired competency, it can be supplied through education or through a new team member.

The assembled team collaborates to establish the building's desired function (proposition 1), which is required prior to development of a decision-based design model (proposition 2). Since the building's energy performance and air quality are emphasized on sustainable projects, the design process can highlight decisions impacting these building attributes. Once the decision-based design process model is developed, key decisions can be evaluated based on their timing in the overall process as well as the sequence in which they are organized (proposition 3). After modelling the design process and analysing key decisions for appropriate timing and sequencing, the information required to make informed decisions can be determined (proposition 4). Finally, each decision and analysis can be evaluated for critical individual

competencies required to effectively make a decision and reach a commitment (proposition 5). These activities continue throughout the design process, aiming for continuous process improvement.

CONCLUSIONS

Conclusions drawn based on this research should consider the limited number and location of case studies as well as the absence of rival theories in the case study data collection phase. Rival theories were considered during the determination of event requirements for proposition validation; however, during the case study event identification, rival theories were not actively sought.

CONTRIBUTIONS

The primary contribution of this research is the development of an evaluation method for the sustainable building design process. Responding to the absence of a continuous, process-based method in the case study projects, the DPEM[SB] was developed as one option to help project teams and researchers plan the design process in an effort to improve the process through added value. The five steps of the DPEM[SB] are:

- determine the building's desired function and form the team
- develop a decision-based design model
- evaluate key decisions for value added based on timing and sequencing
- identify information considerations needed for key decisions; and
- identify competency requirements for process implementation.

This five-step evaluation process is iterative and includes multiple reviews during the design process. These multiple reviews contribute to an approach that aims for continuous value improvement of the design process for sustainable buildings.

FUTURE RESEARCH

This research developed a context within which the design process for sustainable buildings can be evaluated and discussed. Many different design process attributes (e.g. decisions, information, commitments and competencies) have been considered, but not fully explored. This research has highlighted the following opportunities for further research:

- Conduct a broad study of the design process for sustainable buildings. Specific issues to address include the identification of decisions and information with the greatest impact on sustainable building project outcomes. The continuing growth of the worldwide sustainable building market will provide larger quantities of data that could enable statistically significant testing.
- Measure the impact of implementing the DPEM[SB] approach. The value and waste insights developed in the current research can act as guidance in quantifying the impact of the DPEM[SB].
- Investigate the relationship between project outcomes and the presence of team competencies.

AUTHOR CONTACT DETAILS

Christopher S. Magent: Alexander Building Construction LLC, 2545 North Atherton St, State College, PA 16803, US.

Sinem Korkmaz (corresponding author): Michigan State University, 201D Human Ecology, East Lansing, MI 48824, US. Tel: +1 517 353-3252, fax: +1 517 432 8108, e-mail: korkmaz@msu.edu

Leidy E. Klotz: Clemson University, 208 Lowry Hall, Clemson SC 29634, US.

David R. Riley: The Pennsylvania State University, 104 Engineering Unit A, University Park, PA 16802, US.

REFERENCES

Austin, S., 2001, 'Mapping the conceptual design activity of interdisciplinary teams', in *Design Studies*, 22(3), 211–232.

Babbie, E., 2004, *The Practice of Social Research*, Belmont, Wadsworth/ Thomson Learning Publications.

Birmingham, R., Cleland, G., Driver, R. and Maffin, D., 1997, *Understanding Engineering Design: Context, Theory, and Practice*, London, New York, Prentice Hall.

BRE (Building Research Establishment), 2007, *BRE Environmental Assessment Method*, www.breeam.org/ (accessed 27 November 2008).

BHKR (Burt Hill Kosar Rittelmann Associates), 2003, *Achieving Architectural and Engineering Collaboration in Building Design*, White Paper, Butler, PA.

Cross, N., 1994, *Engineering Design Methods*, 2nd edn, Chichester, John Wiley & Sons.

De Wilde, P., Augenbroe, G. and Van Der Voorden, M., 2002, 'Managing the selection of energy saving features in building design', in *Engineering, Construction and Architectural Management*, 9(3), 192–208.

Evbuomwan, N.F.O. and Anuba, C.J., 1998, 'An integrated framework for concurrent life-cycle design and construction', in *Advances in Engineering Software*, 5(7–9), 587–597.

Haviland, D., 1994, *The Architect's Handbook of Professional Practice: 12th Student Edition*, Washington, DC, AIA Press.

Hazelrigg, G.A., 1998, 'A framework for decision-based engineering design', in *Journal of Mechanical Design*, 120(4), 653–658.

Kaatz, E., Root, D., Bowen, P. and Hill, R., 2006, 'Advancing key outcomes of sustainability building assessment', in *Building Research and Information*, 34(4), 308–320.

Kashyap, M., Khalfan, M. and Zianul-Abidin, N., 2003, 'A proposal for achieving sustainability in construction projects through concurrent engineering', in D. Proverbs (ed), *Proceedings of the RICS Foundation*

Construction and Building Research Conference, School of Engineering and the Built Environment, University of Wolverhampton, Wolverhampton, September 2003, London, The RICS Foundation in association with the University of Wolverhampton Press, 127–138.

Korkmaz, S., Riley, D. and Horman, M., 2007, 'Effective indicators for high performance green building delivery', in M. Garvin *et al* (eds), *Proceedings of the 2007 ASCE/CIB Construction Research Congress, Grand Bahama Island, Bahamas, May 2007*, Louisville, Colorado, Academic Event Planners.

Klotz, L., Horman, M. and Bodenschatz, M., 2007, 'A modeling protocol for evaluating green project delivery', in *Journal of Lean Construction*, 3(1), 1–18.

Koskela, L. and Huovila, P., 1997, 'On foundations of concurrent engineering', in C. Anumba and N. Evbuomwan (eds), *Concurrent Engineering in Construction CEC97, The Institution of Structural Engineers, London, July 1997*, 22–32.

Kratzenbach, J. and Smith, J., 2003, *The Wisdom of Teams*, New York, HarperCollins Publishers.

Lapinski, A., Horman, M. and Riley, D., 2006, 'Lean processes for sustainable project delivery', in *ASCE Journal of Construction Engineering and Management*, 132(10), 1083–1091.

Lewis, M., 2004, 'Integrated design for sustainable buildings', in *ASHRAE Journal*, 46(9), S22–S29.

Liker, J.K., 2004, *The Toyota Way*, New York, McGraw Hill.

Luecke, R., 2004, *Creating Teams with an Edge*, Boston, MA, Harvard Business School Publishing Company.

Magent, C., 2005, *A Process and Competency-based Approach to High Performance Building Design*, PhD thesis, Department of Architectural Engineering, Penn State University, PA.

Mistree, F. and Allen, J., 1997, 'Optimization in decision-based design', in A.D. Belegundu and F. Mistree (eds), *Optimization in Industry, Palm Coast, Florida, March 1997*, New York, The American Society of Mechanical Engineers, 1–12.

Neufville, R., 2001, 'Real options: Dealing with uncertainty in systems planning and design', in *Proceedings of the 5th International Conference on Technology Policy and Innovation, Delft, Netherlands, June 2001.*

Raynsford, N., 2000, 'Sustainable construction: The government's role', in *Proceedings of ICE*, 138, November, 16–22.

Reed, W.G., 2003, *The Cost of LEED Green Buildings*, Natural Logic, www.uvm.edu/~rboumans/BillReed_CostOfGreen.pdf (accessed 22 January 2009).

Reed, W. and Eisenberg, D., 2003, 'Regenerative design: Toward the re-integration of human systems with nature', in *A Contribution to the Mayor's Green Building Task Force*, Boston, MA, www.bostonre developmentauthority.org/gbtf/documents/RegenerativeDesign-Reed03-12-18.PDF

Reed, W. and Gordon, E., 2000, *The Integrated (Sustainable/Whole System) Design and Building Process*, White Paper, Boston, MA.

Riley, D., Magent, C. and Horman, M., 2004, 'Sustainable metrics: A design process model for sustainable buildings', in *Proceedings of the CIB World Building Congress, Toronto, Canada, May 2004.*

Simon, H.A., 1969, *The Sciences of the Artificial*, Cambridge, MA, IT Press.

Tzortzopoulos, P., Sexton, M. and Cooper, R., 2005, 'Process models implementation in the construction industry: A literature synthesis', in *Engineering, Construction and Architectural Management*, 12(5), 470–486.

USGBC (United States Green Building Council), 2008a, *A National Green Building Research Agenda*, www.usgbc.org/ShowFile.aspx?DocumentID=3402 (accessed 27 November 2008).

USGBC (United States Green Building Council), 2008b, *The LEED Rating System*, www.usgbc.org/DisplayPage.aspx?CMSPageID=222 (accessed 27 November 2008).

Womack, J. and Jones, D., 1996, *Lean Thinking: Banish Waste and Create Wealth in your Corporation*, New York, Simon & Schuster.

Yin, R., 2003, *Case Study Research Design and Methods*, London, Sage Publications.

ARTICLE

The Construction Design Manager's Role in Delivering Sustainable Buildings

Frederick T. Mills and Jacqueline Glass

Abstract

The emerging champion of the design process is arguably the design manager, increasingly playing a pivotal role in the delivery of sustainable buildings. This research aimed to assess the ability of construction design managers to integrate sustainability into building design, with particular emphasis on the importance of skills. Data were obtained from an extensive literature review, semi-structured interviews with experienced design managers and a survey of graduate design managers. The research identifies issues with construction design management generally, in addition to barriers to the delivery of sustainable buildings. Outcomes of the research contribute to the emergent dialogue on construction design management with regard to sustainable building design.

■ *Keywords* – Construction projects; design management; project teams; skills; sustainability management

INTRODUCTION

It is widely accepted that the construction industry is best placed to influence sustainable development because its 'end product', the built environment, is the context for the majority of human activity (Cofaigh *et al*, 1999; Wines, 2000; Addis, 2001). Construction has the potential to shape how we live our lives and to encourage/enable us to live them in a sustainable manner; by using fewer finite resources, contributing to the development of social capital and supporting the local economy (Williams and Dair, 2006).

The most important aspect of delivering a sustainable building is sustainable building design (Edwards and Hyett, 2005; Stanton, 2006). During this planning phase, materials and construction methods are specified and the manner in which occupants will go about their lives ultimately determined. Indeed, early consideration of sustainability appears key to realizing a sustainable building (Edwards and Hyett, 2005). Among the various core professionals involved

with these activities is the design manager, who has a quasi-construction management role, but who also oversees the process that ultimately results in a sustainable building.

This research investigates the ability of construction design managers to integrate sustainability objectives into the process they manage, with particular emphasis on the importance of skills. Data were initially obtained through a literature review which identified two themes (skills acquisition and skills improvement) for further investigation. These were explored through a small set of opinion questionnaires and semi-structured interviews that sought the views of construction design managers from graduate up to senior level. Data were then analysed, from which a number of tentative conclusions and recommendations were developed. These have implications for government, contractors and their clients in terms of skills acquisition and improvement, with ramifications for the industry's attitudes towards project management, human resources planning, institutional representation and training.

ARCHITECTURAL ENGINEERING AND DESIGN MANAGEMENT ■ 2009 ■ VOLUME 5 ■ PAGES 75–90
doi:10.3763/aedm.2009.0908 © 2009 Earthscan ISSN: 1745-2007 (print), 1752-7589 (online)

from Routledge

CURRENT CHALLENGES TO THE CONSTRUCTION DESIGN MANAGEMENT FUNCTION

Research has shown that the building design process has become more complex over recent years (Beard *et al*, 1998; Gray and Hughes, 2001; Bibby *et al*, 2003a; Tzortzopoulos and Cooper, 2007). This is due largely to the increased popularity of procurement routes (such as design and build) in which responsibility for managing design falls to some extent within the remit of the principal contractor (Beard *et al*, 1998; Tzortzopoulos and Cooper, 2007), rather than with the architect as in more traditional forms of contract. The popularity of such routes is evidenced easily: the majority of UK government-funded infrastructure projects are now being delivered in this way. Further to this, the increased intricacy of buildings can be considered to be a contributory factor to the complexity of the design process (Bibby *et al*, 2003b; Tzortzopoulos and Cooper, 2007). Gray and Hughes (2001) found that the placing of design responsibility on contractors, a process they have little experience in managing (Bibby *et al*, 2003a), created the need for a greater effort in coordination and increased control over each phase of the building project.

Despite the management of design taking place explicitly within other industrial design and manufacturing sectors for a number of decades (Cooper and Press, 1995), the process has only recently risen to the fore in construction as a profession in its own right (Tzortzopoulos and Cooper, 2007). Indeed, many architectural design practices would have discreetly assumed this role by virtue of more traditional forms of contract up until the introduction of design and build.

Gray and Hughes (2001) believe that design management within the construction context is a function that coordinates the design process to deliver high-quality information, enabling the needs of the design, manufacturing and construction processes to be met, while Emmitt (2007), exploring the role from an architect's perspective, describes the role as an information management or coordination function.

Tzortzopoulos and Cooper (2007) define design management as a managerial practice that is focused on improving design procedures, enabling the development of high-quality products (in this case a building) through effective processes. Construction design management is understood to be the coordination and regulation of the building design process, resulting in the delivery of a high-quality building. That said, the existence of various definitions genuinely reflects the piecemeal, somewhat staggered, implementation of design management in the UK, as reported by Gray and Hughes (2001) and others. Construction organizations have tended to implement design management practices independently of one another, which has led to some significant variations in the role description, procedures and practices. It is for this reason that design managers appear to have a number of roles and responsibilities, together with a wealth of tools to assist them. Hence, as an emerging job role in its own right, design management has an emerging set of skills. To map these, some of the key forms of literature (Gray and Hughes, 2001; CIOB, 2007; Tzortzopoulous and Cooper, 2007) and documentary evidence from three major UK design-build contractors (including one that employs around 2500 construction design managers) have been examined. Tables 1 and 2 illustrate an emergent hierarchy for design management skills, together with a matrix of these against the typical responsibilities of a design manager and a number of core design management tools.

Table 2 indicates that the possession of some skills may be more critical than others which supports the notion of the hierarchy developed in Table 1. This classification can only be regarded as tentative, because some skills that seem key to achieving a particular role/using a particular tool are classed as 'supporting'. Therefore, reaching any consensus regarding the skills that a design manager should possess is not as straightforward as identifying specific roles, responsibilities or tools. Tzortzopoulos and Cooper (2007) state that difficulties in understanding a construction design manager's skills relate to the poor understanding of their role; thus any research or attempt to reach a consensus on skills is based on the unstable foundation of a poorly defined profession for whom the daily operating parameters are rather vague. Indeed, their research discovered that the majority of construction design managers appeared to have a poor understanding and knowledge about the design process itself.

TABLE 1 An emergent hierarchy of skills for design management

LEVEL 1 – PRIMARY SKILLS
(proposed by three sources or more)
Comprehension
Communication
Leadership
Teamwork
Planning, coordination and organization
Analytical
LEVEL 2 – SECONDARY SKILLS
(proposed by two sources)
Technical knowledge base
Negotiation
IT competence
LEVEL 3 – SUPPORTING SKILLS
(proposed by one source)
Design flair
Understanding of H&S
Project management
Commercial interface
Design procurement
Strong achievement focus
Firm customer focus
Understanding of market in which contractor operates

Gaining an improved understanding of the nature of the building design process delivered through developments in the skills and knowledge of design managers is a recurring theme (e.g. Cooper and Press, 1995; Bibby et al, 2003b; Bibby et al, 2006). It would seem that the rush into design management (by contractors charged with leading the design process) has arguably resulted in a lack of an education system/training on building design procedures. Another issue that appears to be restricting design managers in practice is a lack of power, which Bibby et al (2003b) attribute to their relatively low position in the project team. The job role does not command the authority it needs to be effective in leading and controlling the design process, which both Bibby et al (2003b) and Cooper and Press (1995) believe limits power and influence. These barriers could prevent the delivery of high-quality design and buildings, regardless of project nature and context. This brings us on to the pressing issue of advances to 'standard practice', in this case sustainable building design and construction and appropriate skills for construction design managers

SUSTAINABLE BUILDING DESIGN MANAGEMENT SKILLS FOR CONSTRUCTION DESIGN MANAGERS

As a process, 'sustainable building design' happens prior to 'sustainable construction', delivers a 'sustainable building' and facilitates 'sustainable development'; as explained against the typical stages of a construction project (RIBA, 2007) in Figure 1. This clearly relates the processes to current construction practice.

Much of the literature (e.g. Addis, 2001; Adetunji et al, 2003; Edwards and Hyett, 2005; Williams and Dair, 2006; DTI, 2007) forms a consensus that sustainable building design is essentially the planning of buildings that:

- are constructed in a sustainable manner
- enable occupants to live a sustainable lifestyle; and
- address Brundtland's definition (World Commission on Environment and Development, 1987) and the 'Triple Bottom Line' concept.

The importance of buildings in successful sustainable communities is recognized within Sir John Egan's *Skills for Sustainable Communities* (Egan, 2004), a report commissioned by the UK government (in response to Brundtland, 1987) that examines the necessary skills required to realize sustainable communities and sustainable development therein. Egan (2004) places particular emphasis on the leaders of the delivery process coming together to attain a 'common goal'. It is widely accepted across several forms of literature (Edwards, 1996; Bennett and Crudgington, 2003; Adetunji et al, 2003; Williams and Dair, 2006; Braithwaite, 2007) that the government is the largest and most frequent client for UK construction and as such is empowered to facilitate this change. By demanding sustainable buildings, the government will see sustainability filter down through contractor supply chains to reach the smaller firms.

Research by Bennett and Crudgington (2003), Bender (2003) and others such as Adetunji et al (2003)

TABLE 2 Design management skills vs design management roles/tools

DESIGN MANAGEMENT SKILLS	TOTAL USAGE	Select and brief design team	Develop, agree and manage design programme	Initiate, maintain and document communication	Prepare contractor's proposals and ensure compliance	Provide significant contribution to risk management	Coordinate/agree design activities with those of other departments
Level 1							
COMPREHENSION	7	*	*		*	*	*
COMMUNICATION	9	*		*			*
LEADERSHIP	5	*	*	*			
TEAM WORK	6	*		*	*		
PLANNING, COORDINATION & ORGANISATION	9		*				*
ANALYTICAL	5		*			*	
Level 2							
TECHNICAL KNOWLEDGE BASE	5					*	
NEGOTIATION	6	*	*	*			*
IT COMPETENCE	4		*				
Level 3							
DESIGN FLAIR	1				*		
UNDERSTANDING OF H&S	2						
PROJECT MANAGEMENT SKILLS	5		*				
COMMERCIAL INTERFACE	2						*
DESIGN PROCUREMENT	1						
STRONG ACHIEVEMENT FOCUS	3		*	*			
FIRM CUSTOMER FOCUS	2				*		
UNDERSTANDING OF CONTRACTORS' MARKET	1				*		

DESIGN MANAGEMENT ROLES/RESPONSIBILITIES

Ensure buildability
of design proposals

Liaise with all parties
that are associated
with/affected by design

Manage design
changes, considering
time and cost impact

Identify and advise
on specialist subcontractors

Ensure safe design in-line
with CDM regulations

Monitor and inform
others of design progress

Produce monthly
design reports

DESIGN MANAGEMENT TOOLS

Project programme
and design/
sub-design programmes

Design and project
meetings/workshops

Information
publication
schedule

Works package list

Online/electronic
document
management

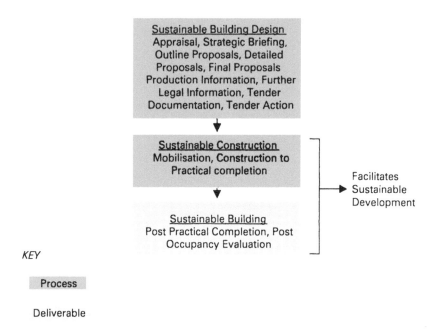

FIGURE 1 Definition of sustainability terms against the typical stages of a construction project (as identified by RIBA (2007))

have found there to be a broad and strong business case for applying sustainable principles to construction, despite scepticism by a minority (Braithwaite, 2007). However, the level of strategic response to sustainability tends to be proportionate to contractor turnover, because sustainability is concerned with long-term survival and this conflicts with aims of many smaller contractors for whom long-term strategic thinking is something of a luxury. Penny (2006) accepts this, but argues that whatever a contractor's size, all will eventually have to operate sustainably to keep up with competitors. However, in practice, sustainable building design remains limited; Williams and Dair (2006) argue that statutory and environmental regulations prescribe minimum standards but that developments often fail to achieve best practice, the most common reason for this being a lack of consideration of sustainability objectives by stakeholders. This could indicate a problem in levels of knowledge, skills, or both and such shortcomings were recognized by Egan (2004). In his review of skills for sustainable communities, he made it clear that sustainability will not be picked up immediately without adequate training for the personnel expected to implement it.

Egan (2004) believes that skilled employees provide a robust infrastructure with which to deliver sustainable buildings and determines several skills necessary to realize sustainable communities. While his review is not exclusively aimed at construction, many of the skills identified are applicable to sustainable building design, as shown in Table 3.

Addis (2001) recognizes that design teams may require guidance during the sustainable design process and highlights areas where assistance may be required. In particular, it would seem critical that construction design managers possess both the skills for successful design management and the appropriate skills for sustainability if they are to play an effective role in delivering a sustainable building. A 'skills deficit' must be identified clearly if it is to be addressed, so Table 4 compares the skills for design management with those necessary for sustainable design and highlights four additional skills required. It is vital that this deficit be addressed if a sustainable building is to be realized.

Indeed, to embed sustainable building design over the long term, it needs to become a standard expectation from stakeholders and the standard

TABLE 3 Applicable generic skills for sustainability (as identified by Egan (2004))

GENERIC SKILL	BEHAVIOURS	
	WAYS OF THINKING	WAYS OF ACTING
Inclusive visioning	Creatively	Entrepreneurial
Project management	Strategic thinking	Can-do mentality
Leadership	Open to change	Cooperation
Breakthrough thinking/brokerage	Awareness of limitations	Able to seek help
Team working	Challenge assumptions	Humility
Determination	Flexible	Committed
Process management	Clear	Respect for diversity
Stakeholder management	Respect for other professionals	Shared sense of purpose
Decisiveness		
Negotiation		
Awareness of the client (incl. how to gain feedback)	None specified	

TABLE 4 Analysis of sustainability vs design management skills

SKILL FOR SUSTAINABILITY (EGAN, 2004)	MATCHING DESIGN MANAGEMENT SKILL*
Inclusive visioning	*No equivalent*
Project management	Project management (3)
Leadership	Leadership (1)
Breakthrough thinking/brokerage	*No equivalent*
Team working	Teamwork (1)
Determination	Strong achievement focus (3)
Process management	*No equivalent*
Stakeholder management	*No equivalent*
Decisiveness	Leadership (1)
Negotiation	Negotiation (2)
Awareness of the client (incl. how to gain feedback)	Firm customer focus (3)

Management skill, its relevant level (from 1–3) is denoted in brackets, as shown in Table 1.* For each design

practice for contractors. This would require a significant step-change in an industry that many describe as highly conservative. Nevertheless, as the driving factors for sustainable building continue to grow (DTI, 2007) the industry will have little choice but to overcome such barriers.

BARRIERS

A number of possible barriers to the construction design manager's role in delivering sustainable building design were established during the review of the literature. Common barriers were found to be a lack of consideration by stakeholders and an unwillingness of the industry to change. In addition, there is a poor understanding of the building design process among design managers, coupled with a lack of authority, which limits their potential influence. These factors are likely to continue to present challenges when dealing with sustainable building design. Key literature from outside the construction sector presents a useful touchstone here. Roome (1998, 2007) takes the view that managing for sustainability is about organizational development and change (in management structures, systems and competencies). Epstein (2008: 199) expands on this, describing organizational learning as the 'new battleground' for sustainability management, dividing core capabilities into:

- skills and knowledge
- physical technical systems (and knowledge)
- managerial systems
- values and norms.

It would appear that possession and proficient use of appropriate skills, both managerial and technical, could be at least part of the answer to overcoming barriers both to effective design management and sustainable building design. The skills issue with which design managers are faced can be essentially categorized as follows – those that they do not possess and must obtain to deliver a sustainable building (i.e. those in the skills deficit) and those that they do possess but must use more effectively. It would be helpful if a framework or model could be developed so that the profession could map its position more clearly with regard to such skills. The research described in the next section therefore focuses on two themes:

- Skills acquisition: To determine ways in which the skills deficit (between those that design managers currently possess and those necessary for sustainable building design) can be addressed.
- Skills improvement: To determine how design managers can use the skills that they already possess to greater effect.

Importantly, the study set out to examine attitudes towards sustainable building design management skills, rather than measuring levels of skills or technical knowledge.

THE RESEARCH APPROACH

The method of research for this study was drawn from the interpretive, qualitative discipline. Opinions, views and feelings of professionals operating in context-specific environments were to be sought and qualitative methods are generally best suited to facilitate this. In this case, a 'multi-method' approach (Saunders et al, 2006) consisting of an opinion questionnaire (seeking to discover people's views) complemented by semi-structured interviews to explore and interrogate those opinions, was used. The overarching aim was to canvass new and experienced

practitioners about the existing levels of skills within the sustainable building design subject and thus to establish the apparent status of the profession, critically identifying areas where skills improvements are needed. Tzortzopoulos and Cooper (2007) made use of semi-structured interviews when conducting a similar study, albeit on a larger scale, into the need for clarity in design management. There, the opinions of professionals were also sought.

A postal/e-mail questionnaire was developed to target graduate design managers (employed by a range of UK-based contractors) who had obtained a BSc degree specializing in design management. This is a relatively new job role and undergraduate degree programmes in the discipline had been established for fewer than seven years at the time of the research. Thus, many design managers (operating at a high level in the industry after lengthy careers) must have entered the role from other backgrounds, such as construction management, project management, architecture or quantity surveying. As a result, they may not have received a curriculum that focuses on design management or sustainable building design, compared with those who have obtained a more specialist degree. Collecting the opinions of graduates in a range of companies provided breadth in the responses. The questionnaire had a total of seven closed-ended and open-ended questions over two sections.

Interviews were also conducted with seven construction design managers employed by a leading UK contractor, with an annual turnover approaching £1 billion in its construction division alone. The company had interests in government-led public infrastructure projects (typically utilizing the design and build procurement route) and, as such, was heavily involved in design management. A tactical approach was taken to collect opinions at a range of seniority levels. In this case, three interviewees were senior design managers and were thus able to speak from a strategic viewpoint, but also in a position of being able to implement/initiate widespread change within the company, where needed.

Bryman and Bell (2003) together with Fellows and Liu (2003) talk of the challenge of finding an 'analytical path' through the rich, qualitative text that is inevitably obtained when conducting semi-structured interviews and surveys. Once the data were obtained, they were analysed in a matrix using a coding technique (Naoum, 1998). This permitted the identification of trends or differences in responses, enabling the extraction of clear results.

A clear limitation of any conclusions subsequently drawn from this research is the modest number of professional opinions sought. The juxtaposition of design management with regard to sustainable building design suggested a number of interesting and dynamic fields where further research could be conducted. A strategic decision was taken by the authors to focus the scope of this research on skills as a means for sustainability implementation and, as such, only a handful of professional viewpoints were obtained.

RESULTS OF THE RESEARCH

This section provides an overview of the results of the analysis, including reference to both the questionnaire and interview data. Of the 22 people approached to take part in the survey, a total of 13 responded. To supplement these responses, seven semi-structured interviews were conducted, five face-to-face and two via the telephone. This gave a total of 20 construction design managers' views. In accordance with the objectives of the research, the results are discussed in two sections: skills acquisition and skills improvement.

THEME 1: SKILLS ACQUISITION

Each individual was invited to discuss how equipped they felt to lead the design of a sustainable building in order to obtain a 'snap-shot' of how prepared the design managers believed themselves to be. The majority felt well equipped, which was widely attributed to experience, twinned (in some cases) with an interest or knowledge of the subject. Graduate design managers referred to the exhaustive education they had received during their studies at university. Interestingly, two schools of thought formed; those who felt that this fully equipped them and those who felt that, due to a lack of experience, this training only equipped them partially. Some went further, suggesting that they felt more able than their superiors to take a lead on sustainable building design, which was attributed to a lack of awareness among their (generally older) colleagues.

Respondents where then asked which skills they felt were necessary for managing/leading the design of a sustainable building. This sought to open a dialogue on skills acquisition, confirm the skills deficit and identify any further skills thought to be relevant by those operating in practice. This question evoked similar responses – knowledge, understanding, awareness, passion, leadership, experience and communication were all mentioned a number of times (Figure 2), although there was very little consensus among the respondents. Some placed sole emphasis on awareness, understanding and/or knowledge, while others felt that 'communication and leadership are a must'.

Finally, respondents were asked how design managers across their respective companies could acquire those skills they had deemed necessary. All respondents mentioned some form of training in their answer with suggestions ranging from compulsory seminars to university degrees. The extent of training required, together with opinion on those most in need of the training seems to be clouded: one graduate spoke of training new employees, while another suggested that senior design managers would benefit from additional guidance. Some respondents emphasized the advantage of experience in order to acquire the necessary skills. One suggested that they can play a more active role on site where an appreciation of sustainable construction processes and the buildability of sustainable design solutions could be fostered.

THEME 2: SKILLS IMPROVEMENT

Respondents were asked to provide their definition of construction design management. This aimed to explore the poor understanding of the profession (even varying between design managers) and its poor theoretical foundation. As perhaps was to be anticipated, various definitions were provided and this supports the findings of Tzortzopoulous and Cooper (2007), Bibby et al (2003a, 2006) and the work of Gray and Hughes (2001). However, several terms were commonly used indicating a broad level of consensus among the respondents. The phrases 'coordination', 'management of design' and 'managing the flow of information' all appeared frequently. Despite this, no two respondents gave an

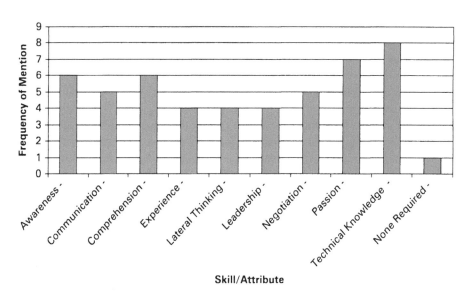

FIGURE 2 Skills felt necessary for managing/leading the design of a sustainable building

identical definition and the extent of a design manager's involvement in the design process appeared to be perceived differently; some used terms such as 'assist', 'guide' or 'support' while others chose words such as 'lead', 'direct' or 'drive', confirming the differences in perception of roles. During the interviews, the apparent confusion over the role of a design manager (suggested by Tzortzopoulos and Cooper, 2007) was evident: one respondent believed the profession comprised 'more management than design' and a senior design manager stated that it entailed 'more design than management' – curiously both respondents were working for the same contractor and in the same office (albeit on different projects). One attributed this to the 'different breeds' of design manager. In fact, several acknowledged that personnel have entered the profession from differing backgrounds and each of these, it was felt, placed emphasis on different aspects of the role. Another respondent stated that each 'breed' of design manager creates a set of procedures that 'works for them' – a somewhat counter-intuitive benefit from having a poorly defined role. Indeed, several interviewees felt that the lack of strict parameters to their role was advantageous. It

would appear that the flexibility afforded enables those from varying backgrounds to adopt an approach that they are comfortable with. This view was not shared by the graduates.

The next question presented two important issues and asked respondents to discuss the extent to which they agreed with/related to them. The first concerned a poor understanding of the design process among design managers, and the second, a lack of authority which limits the potential influence of a design manager. Here, an acknowledgement/ opinion of the skills issue was sought in order to gain insight into the perceived problems and to ignite discussion on skills improvement. The first issue (regarding the design process) received a mixed response as clearly demonstrated in Figure 3. Some felt that this *was* the case, supporting the findings of Bibby *et al* (2003b) and suggesting that the young, inexperienced profession may be the cause. Those agreeing emphasized how important they felt it was that this be resolved. Those disagreeing did so quite strongly. Many insisted that construction design managers were 'very knowledgeable' and that it would be impossible for them to function effectively if they did not understand design. Despite the

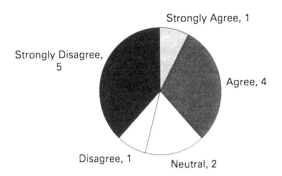

FIGURE 3 Respondent reactions to the finding/notion that construction design managers have a poor understanding of the building design process

differences of opinion, both acknowledged the damaging consequences that a poor understanding of the building design process could have/is having.

The second issue (regarding a lack of authority) was widely acknowledged. However, two categories of opinion developed with regard to how substantial this was and how straightforward it is to overcome. Graduates viewed a lack of authority as a more significant barrier to their performance than their more senior colleagues, who stated that it was a case of proving oneself helpful. Indeed, some interviewees believed the issue to be present but to varying degrees.

The age of the profession, in combination with new procurement routes, was attributed by the respondents in this study to a construction design manager's lack of authority. Not only are many personnel young and hence at an instant disadvantage to their peers, but it was felt that the industry failed to understand both what design management is, and its purpose. Intriguingly, one interviewee pointed out that authority does not automatically earn a construction design manager the respect of his/her peers. Rather, the level of respect a person commands can often be a more formidable force than their level of authority.

Finally, each individual was encouraged to suggest some specific ideas for improvement. The need to raise awareness across the industry regarding the role of the construction design manager in delivering a sustainable building was made explicit, as was the need for a company-wide set of procedures to adhere to. In addition, senior representation of construction design management was called for; a person to champion and lead construction design managers within their own company. Ideally, this person should be experienced and fully aware of the problems faced.

DISCUSSION

SKILLS ACQUISITION

It would be plausible to suggest that any skills deficit could be addressed through various forms of training. However, an emergent theme from the skills acquisition research was that those managing sustainable building design should be equipped with more than a set of skills. Respondents identified many attributes or 'qualities' that a person must possess that are seemingly unobtainable through training or other qualifications, e.g. long-term experience in the industry or, more specifically, experience of delivering sustainable buildings. Further, it appears that if sustainable design is to be realized, it must be forced across the barriers of construction cost, programme time and risk, by someone with a passion for sustainability. This seems more intense than Egan's 'determination' skill (Egan, 2004); it is a deep-rooted belief in an individual of the benefits and urgency of sustainable building design. Respondents felt that this person must act as a role model, encouraging others to follow suit and initiate a 'movement' rather than a policy or contractual agreement which may prescribe merely the minimum standard and become sidelined in later stages. It seems unlikely that the construction industry (which is resistant to change) (Adetunji *et al*, 2003) will adopt sustainable design, unless the benefits are made explicit and proven. Finally, it was felt that those leading design must earn respect as a valued contributor from their peers, independently of any level of authority synonymous with their position.

A new, emergent model for understanding the range of sustainable building design management skills that may need to be acquired is presented in Figure 4. These skills (determined through literature) have been supplemented with those mentioned frequently by the respondents. Figure 4 builds on the management research of Roome (1998, 2007) and Epstein (2008), but most importantly expands the work of Egan (2004) to make these findings explicitly relevant to the construction design managers' role in delivering sustainable buildings.

It is evident that neither of the small demographic samples taken possessed all the qualities illustrated in Figure 4. Graduate construction design managers had received intense training at university establishing firm technical knowledge, whereas more senior construction design managers could draw on past experience and had seemingly developed advanced skills (namely leadership and negotiation). Each demographic therefore has elements (of Figure 4) to offer and it would seem senseless to dismiss either out of hand. Graduates bring new knowledge to the table; while more experienced professionals have usually developed the skills and respect that facilitate implementation. Given the present condition of the profession (with few individuals trained purely in construction design management), a cross-generational, collaborative effort appears to be

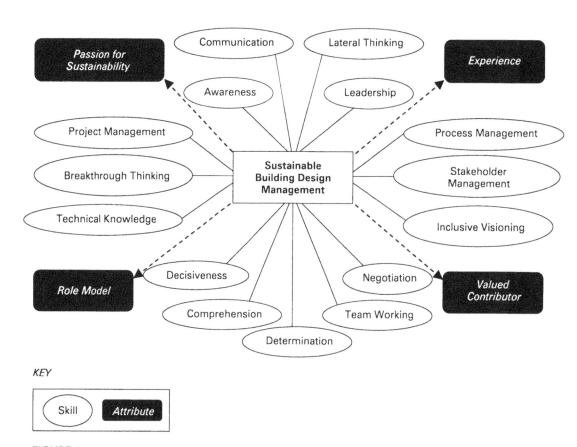

KEY

FIGURE 4 An emergent model for understanding and developing sustainable building design management skills

advisable. This collaboration echoes Egan (2004), who recognizes the need for a common goal across skill disciplines if challenges for which no party is fully experienced or prepared are to be overcome. If the industry is to succeed in delivering sustainable buildings, it is vital that various skill bases are acknowledged and appropriately applied in an integrated manner.

SKILLS IMPROVEMENT

Here widespread resolve is more difficult to achieve. Construction design management has no institutional representation at high level in the industry (such as RIBA or CIOB) and there is a fragmented understanding of design management skills as they exist, both in literature and from the survey results. For example, the practice of personnel entering the job role from different backgrounds seems to have caused problems; some are very knowledgeable about the design process, others are not and perceptions of what the job role entails are quite varied. So, when it comes to construction design managers better using their existing skills, it is difficult to prescribe generically what needs to change. That said, extra training or continuing professional development (CPD) in some of the 'softer' skills areas identified in Figure 4 (such as stakeholder management and negotiation) would be a high-leverage, quick-win solution to enhancing, say, a chartered CIOB or RICS professional with good levels of technical knowledge and site experience.

Beyond this, an interesting question beckons – is design management fundamentally flawed? Gray and Hughes (2001), Lawson (2006) and Frost (1999) all believe that the building design process is highly dynamic, diverse and complex. Is it any wonder then, that attempts to bring order, regiment and control to a procedure which must be fluid and unrestrained to be effective are encountering such difficulties? This is a compelling area for further research.

RECOMMENDATIONS TO STAKEHOLDERS

Directions about how skills for construction design managers in the area of sustainable building design could be improved are proposed here. Table 5 presents a range of actions or recommendations that can be offered to key stakeholders and should be read in conjunction with Figure 4 (which provides the scope of skills development needed). A few key recommendations are explored next.

A new or existing institution should represent construction design management to increase awareness and respect. The role of the construction design manager is pivotal and the relative qualities of design managers at all levels must be acknowledged. Each demographic (graduates through to senior design managers) must be considered on the basis of their strengths rather than their traits (which could only lead to greater fragmentation and a loss of unity). Leading contractors must ensure the job role of design management receives adequate representation so that sound design is valued as an asset to successful projects. Human resource directors and their teams should work closely with construction design managers to initiate this collaboration.

Although bound to an extent by clients' requirements and budgets, it is unacceptable for the construction industry to use these as an excuse to not address sustainability. In addition, contractors must recognize their power to initiate change and not be afraid to adopt new measures in an effort to realize sustainable buildings. Project teams should be wary of the time and cost mindsets that may sideline sustainable intentions and seek to make sustainable design and construction standard practice while the agenda is still a relatively new and upcoming phenomenon.

As an agent of change, construction design managers should champion sustainable building design and the management thereof; they must exude a contagious passion for sustainability that inspires their peers with a similar enthusiasm. This could enhance their ability to influence others (Cooper and Press, 1995) and provide the stimulating new paradigm of embedding sustainability management skills into construction companies, answering calls from authors like Roome (2007) who make an important theoretical link between sustainability management and innovation.

TABLE 5 Summary of recommendations to stakeholders

Action No.	ACTION	LOCAL GOVERNMENT	CENTRAL GOVERNMENT	PRIVATE SECTOR CLIENTS	CONSTRUCTION INDUSTRY IN GENERAL	CONTRACTING COMPANIES	DESIGN MANAGERS
1	Incorporate sustainability into a project's strategic brief	*		*		*	*
2	Educate local government about sustainable building design		*				
3	Consider a new tendering procedure based on whole life cycle costing	*	*		*	*	
4	Consider long-term benefits of sustainable building design	*		*	*		
5	Develop a common goal at national level		*				
6	Reward, praise and share best practice	*	*		*		
7	Establish an institution for the profession of design management		*		*		
8	Ensure design management is represented at high level				*	*	
9	Develop sustainable design and construction as standard practice				*	*	*
10	Initiate a cross-generational skills collaboration among design managers					*	*
11	Champion sustainable building design						*

CONCLUSIONS

In common with previous studies, it remains difficult to identify clarity or depict consensus on many aspects of design management in construction. This review confirmed the belief that it is a developing profession with a lack of clarity concerning its parameters and skills. Data collected from the survey and interviews appeared to compound this portrayal and suggested the need for design management representation at a high level in the industry. A form of institutional body could enhance the occupation's theoretical foundation and unite construction design managers with a common role, earning the profession greater respect. Participants in the research called for better representation within contracting companies. It appears this would raise awareness of a design manager's role, potentially resolving issues with their authority.

It is crucial that barriers to effective design management are addressed if construction design managers are to succeed in exerting influence and delivering sustainable buildings. Chief among the barriers to sustainable development (identified by literature and confirmed by primary research) were a lack of consideration by stakeholders (particularly clients) and an unwillingness of the industry to change. The importance of communicating sustainability in a project's brief was clearly expressed by the respondents as a means of overcoming this. However, briefing on its own appeared insufficient. This study suggests that possession of the appropriate skills for sustainable design appears crucial in overcoming barriers and proceeding with delivery of sustainable building designs. Exploration of the skills acquisition theme found a number of qualities that the graduate, intermediate and senior design managers interviewed felt were necessary for sustainable design but could not be acquired through training. Rather, attributes such as 'experience' or 'passion for sustainability' are obtained in highly dynamic contexts over a number of years. A complex array of qualities appear necessary, which few individuals could possess in entirety. Modest samples of each demographic suggested the presence of wider skills deficits, presenting the need for a cross-generational collaborative effort.

When addressed in harmony, the points raised seem to present a recipe for success; a construction design manager leading the design of a convincing and genuinely sustainable building. But it is here that the challenge lies. An important issue to emerge is the apparent difficulty that construction design managers may face in delivering sustainable buildings. Even if conditions are ideal and a skills collaboration has been orchestrated, it seems that construction design managers must eventually possess many qualities (some of which are seemingly only attainable through long-term experience in the industry). In time (as conditions and individuals are honed and perfected) the construction design manager's role in delivering a sustainable building is likely to become easier.

ACKNOWLEDGEMENTS

The authors wish to thank those design managers (from graduate through to senior level) who participated in this research.

AUTHOR CONTACT DETAILS

Frederick T. Mills: Willmott Dixon Construction Ltd, Munro House, Portsmouth Road, Cobham, Surrey, KT11 1TF, UK. Tel: +44 1932 584700, e-mail: fred.mills@willmottdixon.co.uk
Note: Although Frederick Mills is affiliated with Willmott Dixon Construction, the company was not engaged in the research undertaken.
Jacqueline Glass (corresponding author): Department of Civil and Building Engineering, Loughborough University, Department of Civil and Building, Loughborough University, Loughborough, Leicestershire, LE11 3TU, UK. Tel: +44 1509 228738, fax: +44 1509 223981, e-mail: J.Glass@lboro.ac.uk

REFERENCES

Addis, B., 2001, *Guide C571: Sustainable Construction Procurement: A Guide to Delivering Environmentally Responsible Projects*, London, Construction Industry Research and Information Association (CIRIA).

Adetunji, I., Price, A., Fleming, P. and Kemp, P., 2003, 'Sustainability and the UK construction industry – a review', in *Proceedings of the Institution of Civil Engineers: Engineering Sustainability*, 156(1), 185–199.

Beard, J., Loulakis, M. and Wundram, E., 1998, *Design Build: Planning Through Development*, New York, McGraw.

Bender, B., 2003, 'Sustainable development: UK government objectives and the role of business', in *Proceedings of the Institution of Civil Engineers: Engineering Sustainability*, 156(1), 5–6.

Bennett, J. and Crudgington, A., 2003, 'Sustainable development: Recent thinking and practice in The UK', in *Proceedings of the Institution of Civil Engineers: Engineering Sustainability*, 156(1), 27–32.

Bibby, L., Austin, S. and Bouchlaghem, D., 2003a, 'Defining an improvement plan to address design management practices within a UK construction company', in *International Journal of IT in Architecture, Engineering and Construction*, 1(1), 57–66.

Bibby, L., Austin, S. and Bouchlaghem, D., 2003b, 'Design management in practice: Testing a training initiative to deliver tools and learning', in *Construction Innovation*, London, 3(4), 217–229.

Bibby, L., Austin, S. and Bouchlaghem, D., 2006, 'The impact of a design management training initiative on project performance', in *Engineering Construction and Architectural Management*, 13(1), 7–26.

Bryman, A. and Bell, E., 2003, *Business Research Methods*, Oxford, Oxford University Press.

Braithwaite, P., 2007, 'Improving company performance through sustainability assessment', in *Proceedings of the Institution of Civil Engineers: Engineering Sustainability*, 160(2), 95–103.

CIOB (Chartered Institute of Building), 2007, *The CIOB Education Framework 2007*, www.ciob.org.uk/filegrab/FRAMEWORK.pdf?ref=6 (accessed 19 July 2007).

Cofaigh, E., Fitzgerald, E., McNicholl, A., Alcock, R., Lewis, J., Peltonen, V. and Marucco, A., 1999, *A Green Vitruvius; Principles and Practice of Sustainable Architectural Design*, London, James & James.

Cooper, R. and Press, M., 1995, *The Design Agenda; A Guide to Successful Design Management*, Chichester, John Wiley & Sons.

DTI (Department of Trade and Industry), 2007, *Draft Strategy for Sustainable Construction: A Consultation Paper*, London, DTI.

Edwards, B. and Hyett, P., 2005, *Rough Guide to Sustainability*, 2nd edn, London, RIBA.

Edwards, B., 1996, *Sustainable Architecture; European Directives & Building Design*, 2nd edn, Oxford, Architectural Press.

Egan, J., 2004, *Skills For Sustainable Communities*, London, Office of the Deputy Prime Minister.

Emmitt, S., 2007, *Design Management for Architects*, Oxford, Blackwell Publishing.

Epstein, M., 2008, *Making Sustainability Work*, Sheffield, Greenleaf Publishing.

Fellows, R. and Liu, A., 2003, *Research Methods for Construction*, 2nd edn, Oxford, Blackwell.

Frost, R.B., 1999, 'Why does industry ignore design science?', in *Journal of Engineering Design*, 10(4), 301–304.

Gray, C. and Hughes, W., 2001, *Building Design Management*, Oxford, Butterworth-Heinemann.

Lawson, B., 2006, *How Designers Think: The Design Process Demystified*, 4th edn, Oxford, Architectural Press.

Naoum, S.G., 1998, *Dissertation Research and Writing for Construction Students*, Oxford, Elsevier/Butterworth-Heinemann.

Penny, E., 2006, 'Green is the new black', in *Contract Journal*, 434(6559), 8.

RIBA (Royal Institute of British Architects), 2007, *Outline Plan of Work 2007*, www.snapsurveys.com/learnandearn/ribaoutlineplanofwork 2007.pdf (accessed 1 December 2007).

Roome, N. (ed), 1998, *Sustainability Strategies for Industry*, Washington DC, Island Press.

Roome, N., 2007, 'Developing capabilities and competence for sustainable business management as innovation: A research agenda', in *Journal of Cleaner Production*, 15(1), 38–51.

Saunders, M., Lewis, P. and Thornhill, A., 2006, *Research Methods for Business Students*, 4th edn, Harlow, Financial Times Prentice Hall.

Stanton, J., 2006, 'Get sustainable or lose out', in *Contract Journal*, 434(6601), 10.

Tzortzopoulos, P. and Cooper, R., 2007, 'Design management from a contractor's perspective: The need for clarity', in *Architectural Engineering and Design Management*, 3(1), 17–28.

Williams, K. and Dair, C., 2006, 'What is stopping sustainable building in England: Barriers experienced by stakeholders in delivering sustainable developments', in *Sustainable Development*, 15(2), 135–147.

Wines, J., 2000, *Green Architecture*, Italy, Taschen.

World Commission on Environment and Development (The Brundtland Commission), 1987, *Our Common Future*, Oxford, Oxford University Press.

ARTICLE

The Practice of Sustainable Facilities Management: Design Sentiments and the Knowledge Chasm

Abbas Elmualim, Anna Czwakiel, Roberto Valle, Gordon Ludlow and Sunil Shah

ABSTRACT

The construction industry with its nature of project delivery is very fragmented in terms of the various processes that encompass design, construction, facilities and assets management. Facilities managers are in the forefront of delivering sustainable assets management and hence further the venture for mitigation and adaptation to climate change. A questionnaire survey was conducted to establish perceptions, level of commitment and knowledge chasm in practising sustainable facilities management (FM). This has significant implications for sustainable design management, especially in a fragmented industry. The majority of questionnaire respondents indicated the importance of sustainability for their organization. Many of them stated that they reported on sustainability as part of their organization annual reporting with energy efficiency, recycling and waste reduction as the main concern for them. The overwhelming barrier for implementing sound, sustainable FM is the lack of consensual understanding and focus of individuals and organizations about sustainability. There is a knowledge chasm regarding practical information on delivering sustainable FM. Sustainability information asymmetry in design, construction and FM processes render any sustainable design as a sentiment and mere design aspiration. Skills and training provision, traditionally offered separately to designers and facilities managers, needs to be re-evaluated. Sustainability education and training should be developed to provide effective structures and processes to apply sustainability throughout the construction and FM industries coherently and as common practice.

■ *Keywords* – Sustainability; facilities management; recycling; energy management; climate change

INTRODUCTION

Construction is one of the most important industries in all national economies worldwide (Rodwin, 1987). According to recent figures, the UK construction industry employs more than 1.9 million people with 40% registered as self-employed (Office for National Statistics, 2008). The UK construction industry is dominated by small and medium-sized enterprises with an estimated annual output of more than £83.5 billion in 2007 (Office for National Statistics, 2008).

The sector is highly fragmented with low levels of workload continuity, little interdependence, poor communication and lack of trust. According to Egan (1998), this sector's fragmentation led to the extensive use of subcontracting and prevented the continuity of efficient and effective teamwork. Rodwin (1987) argued that the construction industry is unique in its ability to facilitate development by providing directly for human needs, stimulating investment and generating employment.

Diversity and fragmentation of the industry are due to the various cultural values, processes and interests of the many participating organizations

ARCHITECTURAL ENGINEERING AND DESIGN MANAGEMENT ■ 2009 ■ VOLUME 5 ■ PAGES 91–102
doi:10.3763/aedm.2009.0909 © 2009 Earthscan   ISSN: 1745-2007 (print), 1752-7589 (online)

brought together to deliver a project. The discourse of change towards collaboration in design and construction processes espouses an improvement in contracts, communications and management (Egan, 1998). However, it has been argued that the recommended changes remain an aspiration of the policy makers (Wild, 2002) and academicians (Elmualim, 2007). The inception, design, construction and operation of any construction project require the participation of various agents according to their professional knowledge and experience. By nature, the large number of participants involved in construction and the delivery of projects requires collaboration. However, participants have their own agendas and interests and will mobilize resources, knowledge and practices to meet their own needs. These interests are mainly financial, to achieve a competitive edge over their competitors (Elmualim *et al*, 2006), and have implications for sustainable design management and the overall drive for the sustainability agenda.

Sustainable design is advancing in various industrial sectors due to environmental awareness, rising consumers' ethics, producers' social corporate responsibilities and national and international legislations (Brezet, 1997). Climate change is considered to be the most pressing public policy issue that is driving global corporations to espouse sustainable strategies (Stern, 2006). There are mounting calls for a radical change in the way we design and use products in order to alleviate the damage to our environment (Hawken *et al*, 2000; McDonough and Braungart, 2002) and many aspects of sustainable design have been researched, prototyped and implemented. However, increasingly, case studies of leading buildings are failing to demonstrate any radical change or materialization into good, and common, practice. Hence they are unlikely to contribute towards advancing the sustainability cause (McDonough and Braungart, 2002).

Design management is considered to be the holy grail of sustainability (McDonough and Braungart, 2002) and models and strategies for sustainable design are being developed all the time. Aspects of sustainable design include passive design (whereby the thermal and visual comfort of buildings' users is provided for by manipulating the building form

and fabric), integration of renewables, energy management, water conservation, user participation and stakeholder management (Edwards, 1999; Sassi, 2006; Sayce *et al*, 2004). With the emerging notion of through-life services, Vakili-Ardebili and Boussabaine (2007) suggested a framework for sustainable building design that identifies and measures building values throughout its life. Sassi (2006) called for a contextualization of sustainable design within social and cultural contexts. This is very important as the definition of sustainable design varies with individuals and their cultural and social experiences – hence the diverse, and to a greater extent, polarized meanings and discourses of the terminology of sustainable design (Vakili-Ardebili and Boussabaine, 2007). However, the physical aspect of building still dominates the design and construction with no (or very little) consideration for in-use or facilities management (FM) processes (Kibert *et al*, 2000). There is clearly a need to involve facilities management professions in the design and construction processes to lay bare the sustainability goals (Elmualim *et al*, 2005).

THE SUSTAINABILITY DEBATE

Sustainability is becoming increasingly important for governments, business organizations and the community at large as everyone experiences the consequences of global warming. There is an urgent need to change the way people think and operate. The importance of sustainable development as a tool in the battle against the climate warming is well documented. The European Union as well as the UK government are designing and constantly introducing new legislation that forces the construction industry to achieve improved energy efficiency and reduce carbon emissions. Having said that, the built environment and the construction industry has a serious detrimental impact on the environment. The construction industry accounts for approximately 40% of all resource consumption and is responsible for producing 40% of waste including greenhouse gases (Prasad and Hall, 2004). In the UK, the built environment and especially buildings on their own, use 45% of generated energy to power and maintain them in contrast to 5% used to construct them (CIOB, 2004). The other problem is scarcity of resources and the situation is being aggravated due to developing economies in China and India demanding large amounts of energy and materials. The situation is reflected through the constant rising price of energy due to its limited and finite availability. The affordability and security of the supply is also under question (Norton, 2003). This has a great implication on the sustainable practice of FM. As Hodges (2004) said: 'All these lofty goals to reduce energy consumption and take better care of the environment are of clear benefit to the facility manager. Achieving these goals, however, is easier said than done.' This is due to a discrepancy between the abilities, knowledge, skills and willingness of facilities managers to implement sustainability in their businesses and the fact that they are very often mandated to manage the facilities at the lowest possible, present cost (Shah, 2007). These two desynchronized tendencies prevent the discipline of FM from becoming more sustainable (Hodges, 2004).

The term 'sustainable development' has had many interpretations and there have been numerous attempts to define it. The most cited definition is derived from the report of the World Commission on Environment and Development called *Our Common Future*. The document is also known as the Brundtland Report (1987), as the event was chaired by Gro Harlem Brundtland, the Prime Minister of Norway. The report described sustainable development as 'development that meets the needs of the present without compromising the ability of future generations to meet their own needs'. Sustainability focuses on the strands of economic, social and environmental development as the triple bottom line for sustainable development. The simplicity of the definition of such a complex issue raised an avalanche of criticism (Norton, 2003). Manson (2008) and Norton (2003) argued that the definition lacks purpose and is empty of meaning, but on the other hand, 'it can mean almost anything to almost anyone' (Norton, 2003). However, the concept of sustainable development was generated more than 15 years before the Brundtland Report was created (Norton, 2003). The link between economic development and environmental degradation was first expressed and put on the international agenda in 1972 at the United Nations Conference on the Human Environment in Stockholm. Unfortunately, since then,

there have been very few signs of the implementation of environmental concerns into practice or attempts to minimize the depletion of the ozone and natural resources, or efforts to slow down the process of global warming. Dresner (2002) noted that the environmental situation continued to deteriorate despite international awareness and demand for change.

While economic and environmental factors dominate the debate, the social strand of sustainable development provides a focus on social justice and development, ensuring community participation in decisions for sustainable development and possibly regeneration of neighbourhoods, embracing corporate social responsibility opportunities and possibly even buying Fairtrade catering supplies (Barton *et al*, 2002). There is now an ever-increasing focus on sustainable resource use as advocated by Barton *et al* (2002), where he used the concept of strategic asset management (SAM) as a guiding principle for strategic resource use which includes the principles of ecologically sustainable development for quality of life goals. There is no doubt that achieving sustainability is a priority objective in project delivery: sustainable construction is seen as a vehicle to minimize impacts on the environment, generate minimal waste during the construction process and produce energy-efficient, low-maintenance buildings (Kibert and Wilson, 1999).

Even today there are misunderstandings and misconceptions of sustainable development (Dresner, 2002). The ambiguity of the term is further exacerbated by the use sustainable design or sustainable building terminologies. Indeed there is no easy and agreed definition for sustainable design or sustainable building. The emphasis has always been on promoting eco/green design and energy efficiency in buildings (Sayce *et al*, 2004). Sayce *et al* (2004) articulated that the difficulty in the creation of sustainable buildings lies in the difficulty of a precise definition of the term, the lack of easily recognizable business cases and the fragmentation of interests in the building processes.

The UK government published the Energy White Paper in February 2003. It aimed to put the country on the path to a 60% reduction of carbon dioxide emissions by about 2050, with the majority of the progress made by 2020 (BERR, 2003, 2007a). Other objectives included improving the reliability of the energy supply and the promotion of competitive markets in the UK so that the rate of sustainable economic growth rises together with productivity. In March 2005, the UK government launched a new strategy for sustainable development called *Securing the Future* (DEFRA, 2005). The document presents a set of shared principles that should help to achieve sustainable development. The publication also establishes the priority areas of action shared across the UK as follows:

- sustainable consumption and production
- climate change and energy
- natural resource protection and environmental enhancement
- sustainable communities.

The UK government has further advocated the sustainability agenda by integrating the process of design and procurement processes (Egan, 2004; BERR, 2007b). In terms of sustainable design, all publicly funded projects have to use design quality indictors (DQIs) or equivalent and meet the BREEAM 'excellent' standard. Good design is vital for delivering sustainable buildings. Sustainable design underpins the delivery of sustainable construction but it is vital that sustainable design must contribute to the sustainability triple bottom line of environmental, social and economic sustainability. The BERR (2007b) document marginally alludes to the processes of FM and assets management by using post-occupancy evaluation studies.

Good sustainable design has great effects on human health and well-being (Pearce, 2006). Pearce (2006) also questioned whether construction was sustainable. He further stated that 'how far good design can be credited to the sector [construction], especially the narrowly defined sector, is of course a moot question'.

Indeed all the conjunctions of the design and construction processes and society at large clearly acknowledge the enormity of the task and the polarized discourse in our quest for sustainability. Facilities managers are at the forefront of the quest, if sustainable design and construction is to

materialize. It is vitally important therefore to understand the perceptions of sustainability, the current level of commitment to the sustainability agenda and the barriers perceived by facilities managers in delivering sustainable buildings.

COMPLEXITY OF THE FM INDUSTRY: DEFINITIONS AND SCOPE

Facilities management is one of the fastest-growing professions in the UK. The UK FM market is estimated to be worth £106.3 billion with a forecast annual growth of 2–3% (Shah, 2007). Furthermore, the total turnover of the top 50 FM suppliers has reached £17.7 billion (Moss, 2008). The industry and its market are forecast to develop to include non-core functions such as payroll and IT – activities traditionally not associated with this profession. Brown and Pitt (2001) highlighted the anticipated growth in the airport sector, for example, and the likely impact that sectoral growth will have on the sustainability agenda. The scale of growth in the built environment and the consequential growth of the FM sector is anticipated to be enormous and this will have an impact on environmental sustainability (Brown and Pitt, 2001).

Despite the large market for facilities management, the concept is rather vague. FM as a concept and profession is continually developing. Lord et al (2002) claimed that the term 'originated in the late 1960s to describe the then growing practice of banks outsourcing responsibility for the processing of credit card transactions to specialist providers'. They described the concept as the integration of processes within an organization to maintain and develop the agreed services that support and improve the effectiveness of its primary or core activities. This commonly used definition was first formulated by the European Committee for Standardization (CEN) and later it was formally adopted by the British Institute of Facilities Management (BIFM) (Elmualim et al, 2008). Franklin Becker, one of the FM pioneers, defines the term in the following way: 'FM refers to buildings in use, to the planning, design and management of occupied buildings and their associated building systems, equipment, and furniture to enable and (one hopes) to enhance organizations' ability to meet its business or programmatic objectives' (Becker, 1991).

Despite the myriad definitions of facilities management, the concept largely refers to organizational effectiveness.

Various institutions, professionals and organizations offer different definitions, which reflect the strong relationship and interaction between buildings, services and organizations' core activities. Organizations use buildings, services and assets to create an environment that can enhance the performance of their primary business. The remit of the facilities management industry is very broad and is constantly growing as more activities are tending to be regarded as non-core and included in the facilities management sector. This has great implications for the practice of sustainable facilities management (Elmualim et al, 2008).

SIGNIFICANCE OF SUSTAINABLE FACILITIES MANAGEMENT

The concept of sustainable facilities management has developed in parallel with the overarching concept of sustainable development and the growing appreciation of the scale of predicted climate change (Shah, 2007). Recent extreme weather events such as Hurricane Katrina, flooding in South East Asia and the 2005 and 2007 record hurricane seasons in the Caribbean have illustrated the need to address the threat of global climate change. These experiences have been reinforced by scientific evidence of rising temperatures, the melting of the polar icecaps and glaciers, and revised predictions of more rapid temperature increases and extreme weather conditions. The case for change has been successfully made and the need to balance the three strands of sustainable development – social, economic and environmental – is apparent. It is both fortuitous and timely that the facilities management profession has grasped the agenda for change and is aspiring to develop practical sustainability goals within this rapidly evolving profession (Elmualim et al, 2008). Facilities managers are now at the forefront of organizational behavioural change and in a position to influence the behaviour of individuals working in business, government departments and public services within the facilities they manage. Governments at both national and international level are using regulation to reduce carbon emissions and

manage energy demand. Much of the burden of regulations will need to be picked up by facilities managers at every level (Shah, 2007; Elmualim et al, 2008).

The main aspect of sustainable facilities management is its contribution in the battle against climate change. The Organization for Economic Co-operation and Development (OECD, 2003) highlighted that buildings consume about 32% of the world's resources and that includes 12% of water consumption. The emission of carbon dioxide as a result of such large consumption is enormous and its contribution to the problem of global warming is significant. In this respect, there is great potential for substantial reductions in these patterns and, following that, the substantial reduction of the detrimental effects that they exert on the environment.

Social issues, as well as economic ones, are also under the consideration of sustainable facilities management. The buildings are the environment in which people work and spend 90% of their time indoors, according to the OECD (2003). Creating a healthy environment and good working conditions thus are crucial as they increase the productivity level of the employees and therefore benefit the employers and their businesses. Clements-Croome (2004) explains that it is more expensive to employ people who work in the core business rather than it is to maintain and operate the building and, for this reason, 'spending money on improving the work environment may be the most cost effective way of improving productivity' (Clements-Croome, 2004).

The need for sustainable facilities management, and for skilled facilities managers to carry out this function, is therefore growing and the need to develop new ways of working to meet sustainability criteria is of increasing importance. The concerns now are to meet the challenges of applying sustainable development criteria to the management of facilities integrated within design and construction. This encompasses the life cycle of facilities, from design and construction to disposal, with a strong focus on the operational phase. This operational phase provides a remit to factoring sustainability into maintaining and repairing the physical fabric of the site, obtaining resources based on sustainability criteria, ensuring that this extends through the supply chain, minimizing waste and disposing of it responsibly and reducing energy demand (Elmualim et al, 2008).

Wood (2005) highlights the need to address the existing building stock in achieving sustainability goals. It is this emphasis on the operational phase of buildings that is indeed key to the role of FM, since in the developed world the majority of buildings in current use will remain for the next 50 years, carrying their embodied energy and operational energy requirement into the future (Wood, 1999). The building maintenance and repair market in the UK in 1998 was estimated to be £28 billion (Wood, 2005), and this easily exceeds the £10 billion new build market identified by Wood (1999). This puts into perspective the shared understanding that there has always been a need to manage the physical fabric of buildings, as well as the equipment and furniture within them, and the efficient supply of resources and removal of waste. These functions have existed throughout history, but the complexity of modern society and its increased use of resources mean that there is now more than ever a need for facilities managers of high calibre to meet the needs of business, government and society in the 21st century. Other studies have indicated that more than 75% of facilities managers have responsibility for both the routine upkeep of buildings and their longer-term repair and may operate at a strategic level in this activity (Wood, 1999).

Although the FM profession has been presented with an opportunity to make a real and measurable difference by driving the sustainability agenda forward, it does not at present have easy access to the specialist knowledge, tools and supporting case study material necessary to make this a reality. It is necessary to raise awareness of best practice in the industry and to provide a knowledge portal to share information that will allow professionals to build on their skills in this area. There is a pressing need to research sustainability issues within the facilities management industry, and design and develop the investigative and diagnostic tools required to enable and facilitate the implementation of sustainability measures so as to drive sustainability (Elmualim et al, 2008).

CHALLENGES FOR SUSTAINABLE FM METHODOLOGY

The work presented in this paper forms part of an ongoing research project under the Knowledge Transfer Partnership sponsored by the Technology Strategy Board. The overall aim of the project is to investigate the nature of sustainable facilities management and provide a benefit to the industry and community in the form of best practice guidance. The research is positioned within the interpretative research paradigm (Denzin and Lincoln, 1998). As the researchers are intimately involved in the domain of FM, this research is in accordance with action research principles (Rapoport, 1970; Stringer, 1996) as the objective is to contribute to the understanding of sustainability discourse as well as provide a knowledge portal for practising FM. The research utilizes critical literature reviews, thinking approaches, workshops and questionnaires to shed light on the wider sustainability debate as well as within the FM industry. More workshops and case studies are under way to further develop a knowledge portal for sustainable FM.

The aim was to investigate the nature of sustainable facilities management and its affect on sustainable design management. The objectives are to establish the existing level of perception, understanding and application of sustainable knowledge and practice within the facilities management profession and categorize the key areas of sustainable facilities management where clearer practical tools, information and industry best practice are required. For this purpose, data required were collected through an online survey in the form of self-administered questionnaires to establish the level of sustainable knowledge, commitment and practice within the facilities management industry. The survey was accessed through the BIFM website and made available to subscribing members for a period of one month in November 2006. Prior to the distribution of the questionnaire, various workshops were held to raise awareness about the research project among the BIFM members. The aim of the questionnaire was to obtain data on the existing level of sustainable knowledge and practice within the facilities management industry. It also aimed to establish the areas of facilities management where it is believed that more information is required on the effective implementation of sustainability. The survey takes account of the type of the respondents and attempts to identify their level of commitment or the commitment of the organization they represent to the sustainability agenda.

A pilot questionnaire was tested prior to the commencement of the survey on a small number of potential respondents of the same sample as the final questionnaire. The questionnaire was then presented and discussed in a focus group workshop by a project steering committee comprising 12 practising facilities managers and one academic. The questionnaire was concise, consisting of just eight questions, and used both closed and open question formats. The questions were formatted as follows:

- Question 1 was open ended: 'What does sustainability mean to you and how might you achieve it?'
- This was followed by question 2 which was closed ended: 'Is making your organization more sustainable a key objective for you within the next 12 months?', with the answer options given as 'yes' or 'no'.
- Question 3 was also open ended following on from question 2: 'How might you achieve this?'
- Question 4 sought to elicit information regarding the background of the respondents, phrased as: 'Which of the following options most clearly resembles your involvement with FM?' The options given for this question were: 'in-house FM', 'FM service provider', 'product supplier' and 'other (please specify)'.
- Questions 5 and 6 were closed questions to find information regarding the commitment of respondents' organizations to the sustainability agenda. Question 5 was: 'Does sustainability feature as an objective within your organization's corporate plan?' And question 6 asked: 'Was sustainability reported upon within your organization last annual report?'
- The purpose of question 7 was to identify the knowledge chasm within FM practice. The question was designed in two parts: the first part was 'I/my organization find clearer, practical tools, information and industry best practice useful in

the following areas'. The respondents were given 17 areas identified in the pilot questionnaire and workshops with five options in each category where the respondents can choose from 'strongly agree', 'agree', 'don't know', 'disagree' and 'strongly disagree'. The second part was an open ended box where the respondents were posed with the question: 'In addition to the above. I/my organization believe the following should also be included'.

- Question 8 was to find out whether the respondents would be interested in attending workshops and whether their organization could provide case study material as part of the development of the knowledge portal.

The results presented in this paper are based on the views of 92 respondents. Data collected provided information on the current state of theoretical and practical knowledge among the professionals and identified the knowledge chasm in sustainable FM practice and its implications on sustainable design management. Despite the wide circulation of the survey, respondents may be seen as a fairly focused group with a great interest in the sustainability agenda. Results may therefore depict a best-case scenario, e.g. the level of understanding demonstrated could be higher among respondents compared with non-respondents. However, this methodological stance was not corroborated by the results of the survey as seen in the variation of the understanding of the term sustainability. The results showed that respondents' perception and practice of sustainability had a predominantly environmental focus, suggesting a distinct knowledge gap surrounding the general theory and practice of sustainability in its broadest sense.

PERCEPTIONS AND COMMITMENT TO SUSTAINABILITY

In the classification of respondents by sector, 39 (almost 43%) of all participants of the survey were facilities management service providers, while 33 (nearly 36%) of all respondents were involved in in-house facilities management. Only three respondents (3%) were product suppliers and 17 respondents (nearly 19%) were from a different background. This last group of respondents was comprised of property managers, FM consultants, building maintenance managers or company directors who could not be readily categorized otherwise.

Sustainability is characterized by three strands – environmental, economic and social – and these were taken as the point of reference in assessing the level of theoretical knowledge, with respondents being required to indicate understanding in their answers. For question 1, the responses were very varied as can be seen in Figure 1. More than 23 participants in the survey (25% of the respondents) showed no understanding of the concept, while 63 respondents (68% of the respondents) recognized environmental aspects of the term. The respondents provided some terms describing their understanding of the three bottom lines of sustainability. For instance, environmental concepts included 'protection of the environment', 'waste reduction', 'recycling' and 'limits of natural resources'. Key concepts for social issues included 'corporate social responsibility' and 'personal development'. For economic issues, concepts such as 'business growth' or 'provision of goods and services' were articulated. Among the responses, only one individual quoted Brundtland's definition of sustainable development. The survey results pointed towards a general lack of understanding concerning the holistic nature of sustainability.

Some definitions of the concept of sustainability provided by a selection of the respondents were:

- 'Minimizing the impact of our lives on the planet and its resources. It also means a public corporate responsibility to protect the environment and its resources.'
- 'The efficient use of finite resources involving waste minimization, reuse, recycling and procurement.'
- 'Sustainability is a holistic approach to development. Sustainability is about common sense in terms of the economy, environment and society, and acting responsibly.'
- 'Acting responsibly in what we do so as to make a positive contribution to the communities in which we operate.'
- 'To continuously provide the best services and levels of support with the minimum environmental impact.'
- 'Providing commercial growth and business solutions, without leaving a legacy for future generations.'

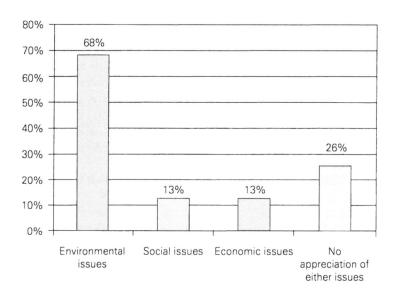

FIGURE 1 Breakdown of respondents' answers which encompassed the three strands of sustainability

- 'Company mission and personal involvement on different levels.'
- 'Being able to maintain a level of service that is acceptable and expected.'
- 'Considering the environmental, social and financial implications of everything I do. Actively minimizing my negative impact on the world and positively enhancing the lives of others through responsible management and purchasing.'

In response to question 2, the majority of the participants, 79 respondents (nearly 86% of the total), indicated that it was important for them to make their organization more sustainable within the next year. With regard to how they may achieve this, a wide range of responses were received. Most of the respondents highlighted challenges such as energy efficiency, recycling and waste reduction in response to question 3. These goals relate to the respondents' interpretation of the environmental aspects as a major part of their concept of sustainability. Below are some of the ideas mentioned by selected respondents to the questionnaire in their answers to question 3 on how to achieve sustainability:

- 'More efficient use of energy, reduction in production of waste, review buying policies to ensure that we are using, where possible, products from renewable resources'.
- 'Better utilities management (electric, water and gas), minimizing waste and more recycling of waste, improvements in green travel alternatives.'
- 'Through the introduction of an environmental policy, the education of staff, focused procurement processes and energy management.'
- 'Selection of an energy supplier with a green resource.'
- 'We believe the best way to make our organization sustainable is to increase our staff awareness of the issues and good practice.'
- 'Develop travel plans, reduce water consumption, reduce energy consumption, reduce–reuse–recycle and raise awareness.'
- In an interesting take on the level of commitment needed to advance the sustainability agenda, a respondent articulated that 'there is a drive to

pursue this approach to business from the main board down to the lower grades of employee and corporate plans are in place to ensure we, and our third party service providers, do all we can to achieve both our corporate and legislative requirements'.

In response to question 5, 70 respondents (76%) reported that sustainability is incorporated as an objective within their organization's corporate plan. For this reason, sustainability can be assumed to be an important issue for the participating organizations and individuals within them. Interestingly in response to question 6, only 43 respondents, just under 47% of all respondents, confirmed that sustainability was reported as an item in their organization's last annual report. Correspondingly, 40 respondents, amounting to nearly 44% of the respondents who answered the question, confirmed that sustainability was not reported upon. When compared with responses to the question regarding the commitment of organizations evidenced by the inclusion of sustainability in the organization's corporate plan, it is clear that a discrepancy exists. Indeed, a commitment through a business plan does not appear to be followed by actions as reporting and auditing.

THE KNOWLEDGE CHASM IN SUSTAINABLE FM PRACTICE

In terms of the identification of the knowledge chasm within the sustainable FM practice (question 7), responses were categorized and prioritized in a list of 17 areas where information and good practice is needed, as can be seen in Figure 2. The top three issues included waste management and recycling, energy management, and specification of sustainable products and services. Other issues included:

- developing a sustainability policy for my organization
- constructing a business proposal incorporating social, environmental and economic benefits
- ethical purchasing
- carbon footprinting
- specification of sustainable M&E
- occupancy satisfaction
- biodiversity

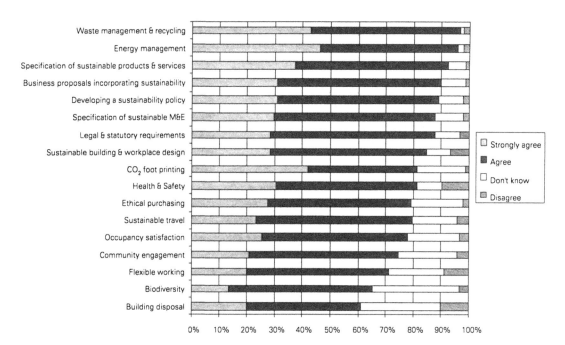

FIGURE 2 Breakdown of the knowledge chasm in sustainable FM practice

- community engagement
- sustainable building and workplace design
- sustainable travel
- legal and statutory requirements
- health and safety
- flexible working
- building disposal.

Many respondents further suggested knowledge gap areas in water management, recycling, and training and education. Other general concerns and thoughts included robustness of information. According to one respondent: 'Information distributed must be correct and not the ramblings of some uninformed environmental weirdo. Misinformation can do more damage than good for the cause of sustainable resources.'

The identification of the knowledge chasm in the practice of sustainable facilities management denoted the information asymmetry between design, and facilities and assets management. Such a wide knowledge chasm further renders any sustainable design management aspiration and practical attempts as mere design sentiments. The

sustainability agenda in the built environment needs to be re-evaluated particularly from a pedagogical perspective. For the sustainability discourse to materialize, there is a pressing requirement for an integrated approach to design, construction and FM. Training and educational structures and curricula require comprehensive evaluation to reflect an integrative approach to sustainable design, construction and FM to encompass the built environment as a whole.

CONCLUSIONS

The construction and facilities management industries have greater leverage and are well positioned to lead humanity's quest for sustainability. With the rising legislative requirements and targets to tackle global warming, construction and FM are in the forefront of delivering the sustainability targets. Theoretical and practical tools for sustainable design management are continually developing. However, little impact is perceived due to the exclusion of FM from these processes, hence rendering any business case for sustainable design as mere sentiments with no bearing effect on sustainability.

The questionnaire survey was conducted to identify perceptions and understanding of the term sustainability, and also to identify the knowledge chasm in the practice of sustainable FM. The results showed that there are challenges in the understanding and building of a consensual definition of sustainability. The existence of a polarized understanding of the term together with materialized interests of various individuals and organizations will remain a barrier for achieving sustainability.

Research findings indicated that the majority of respondents considered the sustainability agenda as important to them and their organizations. Furthermore, the majority stated that sustainability is incorporated as an objective within their organization's corporate plan. Hence, sustainability is seen to be an important issue for the participating organizations and individuals within FM. The respondents indicated that information and practical guidance on a wide range of issues would be helpful in contributing to the implementation of sustainable FM. These issues included energy management, waste management and recycling, developing a sustainability policy for organizations, constructing a business proposal incorporating social, environmental and economic benefits, waste management and recycling, energy management and the specification of sustainable products and services, to mention a few.

The lack or limited understanding of the key concept of sustainability and a lack of practical knowledge have severe implications for the effectiveness of sustainable practice in design management and the FM industry. The fragmentation of the design and construction processes and the asymmetry of practical information during the design and operation of facilities will impede acceleration of the sustainability agenda forward. The lack of information and understanding of the key issues hinders the implementation of sustainability in the FM industry, which could undermine the benefits of sustainable design. Indeed, the knowledge chasm within the sustainable FM practice and sustainable design sentiments has a great implication on education and training not only for FM professionals but also for design and construction management professionals. Coherent and effective structures and educational processes for integrated design,

construction and FM are needed, if sustainability discourses are to be materialized.

ACKNOWLEDGMENT

The authors would like to acknowledge the contribution and the support of the British Institute of Facilities Management, Jacobs plc and Kinnarps (UK) Ltd.

AUTHOR CONTACT DETAILS

Abbas Elmualim (corresponding author): ICRC, The School of Construction Management and Engineering, University of Reading, UK. Tel: +44 118 378 7344, fax: +44 118 931 3856, e-mail: a.a.elmualim@reading.ac.uk

Anna Czwakiel: ICRC, The School of Construction Management and Engineering, University of Reading, UK. e-mail: kcu05akc@reading.ac.uk

Roberto Valle: ICRC, The School of Construction Management and Engineering, University of Reading, UK, and British Institute of Facilities Management (BIFM). e-mail: r.c.valle@reading.ac.uk

Gordon Ludlow: British Institute of Facilities Management (BIFM), UK. e-mail: gpludlow@yahoo.com

Sunil Shah: Jacobs, Reading, UK. e-mail: s.shah@jacob.com

REFERENCES

Barton, R., Jones, D. and Gilbert, D., 2002, 'Strategic asset management incorporating ecologically sustainable development', in *Journal of Facilities Management*, 1(1), 70–84.

Becker, F., 1991, *The Total Workplace – Facilities Management and the Elastic Organisation*, New York, Van Nostrand Reinhold.

BERR (Department for Business, Enterprise and Regulatory Reform), 2003, *Energy White Paper 2003: Our Energy Future – Creating a Low Carbon, Economy*, www.berr.gov.uk/files/file10719.pdf (accessed 15 January 2008).

BERR (Department for Business, Enterprise and Regulatory Reform), 2007a, *Energy Consumption in the UK*, www.berr.gov.uk/files/file11250.pdf (accessed 5 December 2007).

BERR (Department for Business, Enterprise and Regulatory Reform), 2007b, *Strategy for Sustainable Construction*, London, BERR.

Brezet, H., 1997, 'Dynamics in eco-design practice', in *UNEP Industry and Environment*, 20(1–2), 21–24.

Brown, A.W. and Pitt, M.R., 2001, 'Measuring the facilities management influence in delivering sustainable airport development and expansion', in *Facilities*, 19(5–6), 222–232.

Brundtland, G., 1987, *Our Common Future: The World Commission on Environment and Development*, www.re-set.it/documenti/1000/1800/1850/1856/brundtland_reportpdf.pdf (accessed 12 June 2007).

CIOB, 2004, *Sustainability and Construction*, Ascot, Chartered Institute of Building.

Clements-Croome, D., 2004, *Intelligent Buildings: Design, Management and Operation*, London, Thomas Telford.

DEFRA (Department for Environment, Food and Rural Affairs), 2005, *Social Enterprise Securing the Future*, produced for DEFRA by *Social Enterprise Magazine*, www.socialenterprisemag.co.uk/upload/documents/document10.pdf (accessed 19 March 2009).

Denzin, N. and Lincoln, Y.S., 1998, *The Landscape of Qualitative Research: Theories and Issues*, London, Sage Publications.

Dresner, S., 2002, *The Principles of Sustainability*, London, Earthscan Publications.

Edwards, B., 1999, *Sustainable Architecture: European Directives and Building Design*, 2nd edn, Oxford, Architectural Press.

Egan, J., 1998, *Rethinking Construction*, London, DETR/Stationery Office.

Egan, J., 2004, *The Egan Review: Skills for Sustainable Communities*, London, Office of the Deputy Prime Minister, RIBA Enterprise.

Elmualim, A.A., 2007, 'Psycho-social approach for discerning the dichotomy of competition and collaboration to advance mutualistic relationships in construction', in W. Hughes (ed), *CME25: Construction Management and Economics: Past, Present and the Future, University of Reading, UK, July 2007*, Spon Press.

Elmualim, A.A., Fernie, S. and Green, S., 2005, 'Harnessing the role of FM in the design processes through post-occupancy evaluation studies', in *Combining Forces, Advancing Facilities Management and Construction through Innovation, Helsinki, Finland, June 2005*, Finland, VTT, 548–559.

Elmualim, A.A., Green, S.D., Larsen, G. and Kao, C.C., 2006, 'The discourse of construction competitiveness: Material consequences and localised resistance', in M. Dulaimi (ed), *Joint International Conference on Construction Culture, Innovation and Management (CCIM), Dubai, November 2006*, The British University in Dubai, UAE and CIB.

Elmualim, A.A., Czwakiel, A., Valle, C.R., Ludlow, G. and Shah, S., 2008, 'Barriers for implementing sustainable facilities management', in *World Sustainable Building Conference 2008, Melbourne, Australia, September 2008*, CSIRO, Australia.

Hawken, P., Lovins, A. and Lovins, L.H., 2000, *Natural Capitalism: Creating the Next Industrial Revolution*, New York, Little, Brown and Company.

Hodges, C.P., 2004, 'A facilities manager's approach to sustainability', in *Journal of Facilities Management*, 3(4), 312–324.

Kibert, C.J., Sendizimir, J. and Guy, B., 2000, 'Construction ecology and metabolism: Natural system analogues for sustainable built environment', in *Construction Management and Economics*, 18(8), 903–916.

Kibert, J.C. and Wilson, A., 1999, *Reshaping the Built Environment: Ecology, Ethics and Economics*, Washington, DC, Island Press.

Lord, A., Lunn, S., Price, I. and Stephenson, P., 2002, *Emergent Behaviour in a New Market: Facilities Management in the UK*, report by Facilities Management Graduate Centre, Sheffield Hallam University, UK.

Manson, N.A., 2008, 'The concept of irreversibility: Its use in the sustainable development and precautionary principle literatures', in *Electronic Journal of Sustainable Development*, 1(1), www.ejsd.org/ public/journal_article/2

McDonough, W. and Braungart, M., 2002, *Cradle to Cradle: Remaking the Way we Make Things*, New York, North Point Press.

Moss, Q.Z., 2008, 'FM market research review: do we really have the "intelligence"?', in *Facilities*, 26(11/12), 454–462.

Norton, B., 2003, *Searching for Sustainability: Interdisciplinary Essays in the Philosophy of Conservation Biology*, Cambridge, UK, Cambridge University Press.

Office for National Statistics, 2008, *Construction Statistics Annual*, London, Office of National Statistics.

OECD (Organization for Economic Co-operation and Development), 2003, *Environmentally Sustainable Buildings: Challenges and Policies*, Paris, OECD.

Pearce, D., 2006, 'Is the construction sector sustainable? definitions and reflections', in *Building Research and Information*, 34(3), 201–207.

Prasad, D. and Hall, M., 2004, *Construction Challenge: Sustainability in Developing Countries*, London, RICS Leading Edge Series.

Rapoport, R.N., 1970, 'Three dilemmas in action research', in *Human Relations*, 23, 499–513.

Rodwin, L., 1987, *Shelter, Settlement and Development*, Nairobi, UNCHS (United Nations Centre for Human Settlement) (Habitat).

Sassi, P., 2006, *Strategies for Sustainable Architecture*, Oxford, UK, Taylor and Francis.

Sayce, S., Walker, A. and McIntosh, A., 2004, *Building Sustainability in the Balance: Promoting Stakeholder Dialogue*, London, EG Books.

Shah, S., 2007, *Sustainable Practice for the Facilities Manager*, Oxford, Blackwell Publishing.

Stern, S.R., 2006, *Stern Review on the Economics of Climate Change*, London, Cabinet Office, HM Treasury, UK.

Stringer, E.T., 1996, *Action Research: A Handbook for Practitioners*, San Francisco, California, Sage Publications.

Vakili-Ardebili, A. and Boussabaine, A.H., 2007, 'Creating value through sustainable building design', in *Architectural Engineering and Design Management*, 3(2), 83–92.

Wild, A., 2002, 'The unmanageability of construction and the theoretical psycho-social dynamics of projects', in *Engineering, Construction and Architectural Management*, 9(4), 345–351.

Wood, B., 1999, 'Intelligent building care', in *Facilities*, 17(5–6), 189–194.

Wood, B., 2005, 'The role of existing buildings in the sustainability agenda', in *Facilities*, 24(1/2), 60–67.

Architectural Engineering and Design Management: An international journal
Bridging the gap between architecture and engineering practice

Influential and far-reaching, *Architectural Engineering and Design Management* (AEDM) provides a unique forum for the dissemination of academic and practical developments related to architectural engineering and building design management. This new international, peer-reviewed journal details the latest cutting-edge research and innovation within the field, spearheading improved efficiency in the construction industry. Informative and accessible, this publication analyses and discusses the integration of the main stages within the process of design and construction and multidisciplinary collaborative working between the different professionals involved.

Ideal for practitioners and academics alike, AEDM examines specific topics on architectural technology, engineering design, building performance and building design management to highlight the interfaces between them and bridge the gap between architectural abstraction and engineering practice. Coverage includes:

- Integration of architectural and engineering design
- Integration of building design and construction
- Building design management; planning and co-ordination, information and knowledge management, value engineering and value management
- Collaborative working and collaborative visualisation in building design
- Architectural technology
- Sustainable architecture
- Building thermal, aural, visual and structural performance
- Education and architectural engineering

NOTES FOR CONTRIBUTORS TO
Architectural Engineering and Design Management

1. SUBMISSION

Authors should submit one copy of the printed manuscript plus one copy on disk or via e-mail. Authors should ensure that the disk corresponds to the final revised version of the manuscript hard copy exactly. Digital files should be named as follows: author's surname – keyword – date of submission e.g. jones collaborative 251204. All authors are asked to submit full contact details for three potential reviewers of their manuscript. Manuscripts should be submitted to:

Professor Dino Bouchlaghem, Department of Civil and Building Engineering, Loughborough University, Leicestershire, LE11 3TU, UK. E-mail: n.m.bouchlaghem@lboro.ac.uk

Authors should keep a copy of the articles and illustrations.

While the Editors, Referees and Publishers will take all possible care of material submitted to the Journal, they cannot be held responsible for the loss of or damage to any material in their possession.

All articles will be peer-reviewed before acceptance. The final decision on acceptance will be made by the appropriate editor.

2. LANGUAGE AND STYLE

Articles should be in English and should be written and arranged in a style that is succinct and easy for readers to understand. Authors who are unable to submit their articles in English should contact the Editors so that any alternatives may be considered. Illustrations should be used to aid the clarity of the article; do not include several versions of similar illustrations, or closely-related diagrams, unless each is making a distinct point.

3. MANUSCRIPT PREPARATION AND LAYOUT

The manuscript should be printed in double-spacing on one side of the paper only with a wide margin on the left hand side. The pages should be numbered consecutively. Headings and subheadings should be used so that the paper is easy to follow.

The first page of the manuscript should contain the full title of the article, the author(s) names without qualifications or titles, and the affiliations and full address of each author. The precise postal address, telephone and fax numbers and email address of the author to whom correspondence should be addressed should also be included.

The second page should contain an abstract of the article and a key word list (5–10 words). The abstract should be no more than 200 words long and should précis the article, giving a clear indication of its conclusions. The maximum length for the entire manuscript is 7000 words.

Tables and Schema

Each of these should be on a separate sheet, at the end of the manuscript, clearly labelled with the name of the main author. Authors should aim to present table data as succinctly as possible and tables should not duplicate data that are available elsewhere in the article.

Symbols, Abbreviations and Conventions

Please use SI (Systeme Internationale) units. Whenever an acronym or abbreviation is used, ensure that it is spelled out in full the first time it appears. Please indicate in the margin any unusual symbols such as Greek letters that are used in the article.

References and Notes

References should be presented in 'author/date' style in the text and collected in alphabetical order at the end of the article. All references in the reference list should appear in the text. Each reference must include full details of the work referred to, including paper or chapter titles and opening and closing page numbers.

JOURNALS

Smith, D.W., Davis, L.M. and Price, B.N., 2004, 'Integrated Decision-Making in Construction', in *Architectural Engineering and Design Management*, 1(1), 64–72.

BOOKS

Price, B.N., 1993, 'Integrated Decision-Making in Construction', in L.M. Davis, (ed.), *Integrated Thinking*, London, James & James (Science Publishers) Ltd, 84–92.

Smith, D.W. and Davis, L.M., 1993 *Integrated Thinking*, London, James & James (Science Publishers) Ltd.

PROCEEDINGS

Smith, D.W., Davis, L.M. and Price, B.N., 1999, 'Integrated Decision-Making in Construction', in *Proceedings of the 8th Annual Meeting of the Society for Architectural Engineering*, London, James & James (Science Publishers) Ltd.

Notes – which should be kept to a minimum – will appear as endnotes. Indicate endnotes with a superscript number in the text, and include the text at the end of the article. Do not use the footnote/endnote commands in word processing software for either references or notes.

Illustrations

Illustrations should be included at the end of the manuscript. Photographs and line drawings should be referred to in the text as Figure 1, Figure 2. etc. Each illustration requires a caption.

Illustrations should be numbered in the order in which they appear. They should be submitted in a form ready for reproduction – no redrawing or re-lettering will be carried out by the Publishers. Each illustration should be clearly marked with the figure number, the name of the main author, and the orientation if it is needed.

Images can be supplied as tiff or jpeg files (not embedded in the text file) or digitally. Images must be at least 300 dpi at 140 mm wide. Note that figures and graphs must be comprehensible in black-and-white – use patterns, not colours, to differentiate sections. If colour is essential, in most cases the additional cost for including colour will be borne by the author. Please use a consistent font and point size in figures, bearing in mind that most figures will be either 7 or 14 cm wide in the printed journal.

4. PROOFS AND OFFPRINTS

The corresponding author will receive proofs for correction; these should be returned to Earthscan within 48 hours of receipt. The corresponding author will be sent a pdf of the published article. If printed offprints are required, these must be ordered prior to publication; an order form will be provided.

5. COPYRIGHT

Submission of an article to the journal is taken to imply that it represents original work, not under consideration for publication elsewhere. Authors will be asked to transfer the copyright of their articles to the publishers. Copyright covers the distribution of the material in all forms including but not limited to reprints, photographic reproductions and microfilm. It is the responsibility of the author(s) of each article to collect any permissions and acknowledgements necessary for the article to be published prior to submission to the Journal.